Hybrid Geographies

Sarah Whatmore

Hybrid Geographies

natures cultures **spaces**

SAGE Publications
London • Thousand Oaks • New Delhi

First published 2002
Reprinted 2006

SAGE Publications Ltd
1 Oliver's Yard
55 City Road
London EC1Y 1SP

SAGE Publications Inc.
2455 Teller Road
Thousand Oaks, California 91320

SAGE Publications India Pvt Ltd
B-42, Panchsheel Enclave
Post Box 4109
New Delhi 110 017

British Library Cataloguing in Publication data
A catalogue record for this book is available from the British Library.

ISBN-10 0-7619-6566-1
ISBN-10 0-7619-6567-X (pbk)
ISBN-13 978-0-7619-6566-4 (hbk)
ISBN-13 978-0-7619-6567-1 (pbk)

Library of Congress catalog card number 2002102286

**for
Denys E. Whatmore**

*You embark; you make the voyage;
you reach port: step ashore, then.* *

* Marcus Aurelius, *Meditations (Book Three)*. Penguin edition, 1964: 55.
Translated by M. Staniforth. Penguin Books, London.

Contents

List of figures and table

Preface and acknowledgements

This book has been some time in the making. I have tried to hold on to some sense of this energetic fabrication in the writing which weaves my journeying in the space/times of this research through all manner of sustained and chance encounters with others whose assistance and company afford quite different tacks. I am pleased to be able to acknowledge at least some of these many debts without divesting myself of any responsibility for what follows. These journeys began in 1993 under the auspices of a Global Environmental Change Fellowship funded by the Economic and Social Research Council (award no. L320273073). An early version of chapter 7 was written during this fellowship, which also supported the research on which chapters 4 and 5 are based. The research for the chapters in section 1 was funded by another ESRC research grant 1996–97 (award no. R000222113), while the award of a Haggett Fellowship in 1999 by the School of Geographical Sciences at the University of Bristol facilitated the research for chapter 6 and enabled me to write a large part of this book.

I am grateful to numerous colleagues and friends for ongoing conversations or specific engagements with versions of one or more of the chapters that follow. In this regard, I would like to thank Kay Anderson; Trevor Barnes; Nick Bingham; Fred Buttel; Noel Castree; Gail Davies; David Demeritt; J.D. Dewsbury; Margaret Fitzsimmons; David Goodman; Kevin Hetherington; Steve Hinchliffe; Jane Jacobs; Owain Jones; Jack Kloppenberg; Doreen Massey; Mike Michael; Marc Mormont; Jon Murdoch; Phil O'Neill; Bronwyn Parry; Andrew Sayer; Pierre Stassart; Lorraine Thorne; Nigel Thrift; and Michael Watts. Two people warrant particular mention. First, Lorraine Thorne who has shared with me many of the passionate curiosities and strange journeys ventured here, not least as co-author of earlier versions of chapters 2 and 3 which she has generously allowed me to re-work for the purposes of this volume. I remember our floor-scale 'back of the envelope' diagramming fondly, and miss it still. Secondly, Nigel Thrift whose polymathic energies and enthusiasms have fostered an intellectual current of experimentation at Bristol that has nourished my work, alongside many others, over the years to which this book owes much. If academic work were always such serious fun.

I am also indebted to several people for contributing various forms of expertise solicited in the production of this book. Simon Godden and Jonathan Tooby at Bristol worked with skill and good humour on the figures and illustrations. The library staff at the Food and Agriculture Organization in Rome and the Australian Parliament in Canberra were endlessly helpful in furnishing my documentary requests. Pepe Esquinas-Alcazar kindly made available his rich records and remembrances of the political career of the Commission for Plant Genetic Resources and guided me through the FAO labyrinth. Dr Brian Johnson at English Nature; Dr Amy Plowman and staff at Paignton Zoo; and Mr Geoff Wilson at the Earthwatch Institute, Oxford, all made themselves and various materials available amidst busy schedules. In addition, a number of academic institutions have been generous hosts of study visits and/or workshops which have enriched my thinking. In particular, I would like to thank faculty and graduate students in the Department of Rural Sociology at the University of Wisconsin, Madison; in Environmental Studies at the University of California, Santa Cruz; in the Department of Geography at the University of Newcastle (NSW); and in SEED (the Société, Environnement et Developpement research group) at the Universitaire Luxembourgeoise, Arlon.

As an editor Robert Rojek at Sage has been a model of patient encouragement throughout the protracted writing of this book. More than this, his commitment to the project has spurred me on at difficult moments. It is also to his talent for making connections that I owe the cover. In this regard, I am much indebted to Caroline Tisdall who generously agreed to the use of one of her series of Beuys/Coyote photographs and to Robert Violette of Violette Editions who made the image available for reproduction. I would also like to thank Claire Roberts at Sage for her patient work in securing reproduction rights to this and other images in the book.

Finally, I must thank Keith who has borne the brunt of my prolonged preoccupations with characteristic forebearance; Tom and Anna who have indulged them during their visits; my parents who have been unfailingly supportive even when their needs have been much greater; Jan and Jo whose love and friendship has kept me going through it all; and Meg and Dillon who have kept me company during the long hours at 'the top of the house'. I dedicate this book to my father, who did not live to see its completion but whose life shapes mine still in so many enduring ways.

The author and publisher wish to thank the following for permission to use copyright material:

The Saarland Museum (Saarbrucken, Germany) for Figure 2.2, Auguet R., 'Cruelty and civilisation'.

The European EEP, for Figure 3.1, 'Current wild population of *loxodonta africana*', 1997.

Paignton Zoo Environmental Park (Paignton, Devon) for Figure 3.2, 'Giving Animals a Home They Deserve and a Place You Enjoy', 1997.

The Earthwatch Institute, Center for Field Research (Massachussetts, USA) for Figures 3.4 and 3.6 from the Earthwatch Annual Expedition Guide.

Andrew Ireland, for Figure 4.1, 'A Perfect Case of Terra Nulltus!', first reproduced in the *Sydney Morning Herald*.

Mike Bowers, for Figure 4.2, courtesy of *The Age*.

John Shakespeare, for Figure 4.4, first reproduced in the *Sydney Morning Herald*.

Sandy Scheltama, for Figure 4.5, courtesy of *The Age*.

ActionAid, for Figure 5.1, reproduced with kind permission.

The Food and Agriculture Organization of the United Nations, (Rome, Italy), for Figure 5.3, 'The twelve megacentres of cultivated plants', and for Figure 5.4, 'The world's major national plant gene banks', both first reproduced in 'Harvesting Nature's Diversity', 1993.

ARCO Seed Company, for Figure 6.1a, 'Our good taste is a product of culture', first reproduced as company advertisement.

Trez, for Figure 6.1b, his cartoon 'Le Secret Alimentaire', first reproduced in *France Soir.*

National Geographic, for Figure 6.3, for permission to reproduce photograph of soya bean taken by Chris Johns.

Greenpeace, for Figure 6.4, for permission to reproduce photograph 'Roundup Ready'.

The Guardian, for Figure 6.5, for use of a photograph of a GM Protest march first reproduced in *The Guardian*.

Colin Wheeler, for Figure 6.6a, his cartoon 'GM Maize'.

David Austin, for Figure 6.6b, his cartoon 'GM Potato'

The author and publisher would also like to acknowledge the original source of material adapted for use in this volume:

Figure 2.1, adapted by the author from C. Scarre, (1995) *The Penguin historical atlas of ancient Rome*, The Penguin Group, New York.

Figure 4.3, adapted by the author from Gelder, K. and Jacobs, J.M. (1998) *Uncanny Australia: Sacredness and Identity in a Postcolonial Nation*, University of Melbourne Press, Melbourne.

Figure 6.2, adapted by the author from Rost et al., (1994), 'Redefining the goals of protein secondary structure prediction', *Journal of Molecular Biology*, 235, 13–26.

Every effort has been made to trace all the copyright holders, but if any have been overlooked, or if any additional information can be given, the publishers will be pleased to make the necessary amendments at the first opportunity.

Introducing Hybrid Geographies

What happens if we begin from the premise not that we know reality because we are separate from it (traditional objectivity), but that we can know the world because we are connected with it?' (Katherine N. Hayles, 1995: 48)

dis-placing nature – the refrain of the 'outside'

Barely a day passes without another story of the hyperbolic inventiveness of the life sciences to complicate the distinctions between human and non-human; social and material; subjects and objects to which we are accustomed. Variously labelled as 'life politics' (Giddens, 1991) or 'bio-sociality' (Rabinow, 1992a), such worldly apprehensions have struggled to make their mark against academic divisions of labour and the viscous terms in which the 'question of nature' has been posed in the social sciences and humanities (see Macnaghten and Urry, 1999). As their forays into the domain of natural sciences have swelled, so a plethora of 'things' has been trespassing into the company of the social unsettling the conduct of its study. Such things exceed both the proliferation of environmental sub-disciplines[1] and the tired theoretical resources of '(social) constructionism' and '(natural) realism' that have greeted them (see Soper, 1995; Demeritt, 1998). They present an unhappy choice. On the one hand, 'post-modern' modes of enquiry in which Nature, having nothing to say for itself, is the always already crafted product of human interpretation, and analysis becomes fixed on the representational practices that make it meaningful (Robertson *et al.*, 1996). On the other, knowledge projects committed in various ways to maintaining a 'crucial distinction . . . between material processes and relations . . . and our understandings of . . . those processes' (Dickens, 1996: 83) in order to sustain the possibility of (and their own pretensions to) exemption from the representational moment.

There is undoubtedly a generous measure of caricature in this embattled depiction of the treatment of Nature/nature in social theory that serves primarily to reaffirm intellectual prejudices and identities, and which is writ large in the so-called science wars (see Gross and Levitt, 1994). Only the most vulgar of 'constructionist' accounts suggest that the world is – to borrow Sheets-Johnstone's evocative phrase – 'the product of an immaculate linguistic conception' (1992: 46). Equally, only the crudest of 'realist' accounts refuse to recognize the contingency of knowledge claims about 'real world' entities and processes. Moreover, accounts that get lumped into these categories are inevitably more diverse than their detractors acknowledge (see, respectively, Benton, 1996; Conley, 1997). But for all their loudly declared hostility, these theoretical encampments are similarly premised on an a priori separation of nature and society. As Bruno Latour has put it:

> Critical explanation always began from the poles and headed toward the middle, which was first the separation point and then the conjunction point for opposing resources. . . . In this way the middle was simultaneously maintained and abolished, recognised and denied, specified and silenced. . . . How? . . . By conceiving every hybrid as a mixture of two pure forms. (1993: 77–8)

Perhaps because geographers have inhabited this 'nature–society' settlement more self-consciously than other disciplines, these (re)turns to the question of nature have a particular resonance. As every undergraduate knows, Geography stakes its identity on attending to 'the interface between social and natural worlds'. In practice, the separateness of these worlds has been intensified by a disciplinary division of labour between 'human' and 'physical' geography, each of which tends to pay more allegiance to the divergent research cultures of the social and natural sciences respectively than to the other.[2] There is a sense, too, in which the life seems to have been sucked out of the worlds that Geography has come to inhabit, at least in its efforts to become a spatial science and in some more 'critical' spatial theorizing (see Fitzsimmons, 1989).[3] In their urge towards the disembodied authority of panoptic knowledge practices, such maps have 'ceased to be places of sensible activity and journeying' (de Certeau, 1988: 129). More significantly, the spatialities in which the ontological separation of nature and society inheres are woven through all manner of scientific, policy, media and everyday practices that enact nature as 'a physical place to which you can go' (Haraway, 1992: 66). As Tim Ingold has observed: 'Something . . . must be wrong somewhere, if the only way to understand our own creative involvement in the world is by first taking ourselves out of it' (1995a: 58).

Human geography is by no means alone in finding itself at an important juncture in its efforts to escape the dialectical vortex of nature–society relations and the environmental refrain of the 'outside' (see Wolfe, 1998).[4]

The 'hybrid geographies' that I embark on here exercise other modes of travelling through the heterogeneous entanglements of social life that refuse the choice between word and world by fleshing out a different conception of fabric-ation, 'not as mere retro-projection of human labour onto an object that is nothing in itself but a sturdier, much more reflexive co-production richly invested within a collective practice' (Latour, 1999a: 274).

Of course hybridity is already freighted in various ways, for example as the 'margin . . . where cultural differences contingently and conflictually touch' in post-colonial studies (Bhabha, 1994: 206), and in agronomy as the bodying forth of human in(ter)ventions in the flesh of plants (Simmonds, 1979), both of which are interrogated during the course of the book. But its energies are enrolled here primarily as a device to negotiate the temptations of the 'one plus one' logic or 'mixture of two pure forms' that Latour warns against above, in journeying between natures and societies; objects and subjects; humans and non-humans and into their excesses.

'Hybrid geographies' allies the business of thinking space (Crang and Thift, 2000) to that of thinking through the body (Kirkby, 1997), in other words to apprehend and practise geography as a craft. This enterprise gestures towards Michel Serres' insistence that 'there is a sense in space before the sense that signifies' (1991: 13) in two ways: by attending simultaneously to the inter-corporeal conduct of human knowing and doing *and* to the affects of a multitude of other 'message-bearers' that make their presence felt in the fabric of social life. To map the lively commotion of these worldly associations is to travel in them, negotiating 'modes of access and ways of orienting ourselves to the concrete world we inhabit' (Bingham and Thrift, 2000: 292). What happens as a consequence of such mappings into knowledge? A preliminary response to the question staged at the outset would be – an upheaval in the binary terms in which the question of nature has been posed and a re-cognition of the intimate, sensible and hectic bonds through which people and plants; devices and creatures; documents and elements take and hold their shape in relation to each other in the fabric-ations of everyday life (Clark, 1997). As the book goes on to explore, this upheaval implicates *geographical* imaginations and practices both in the purifying impulse to fragment living fabrics of association and designate the proper places of 'nature' and 'society', *and* in the promise of its refusal. This is a promise of countenancing the world as an always already inhabited achievement of heterogeneous social encounters where, as Donna Haraway puts it, 'all of the actors are not human and all of the humans are not "us" however defined' (1992: 67).

diagramming – more than human worlds

> A path is always between two points, but the in-between has taken
> on all the consistency and enjoys both an autonomy and a direction
> of its own. (Deleuze and Guattari, 1988: 380)

The heterogeneous conception of social life that I want to flesh out here takes up the 'common emphases on positionality and interaction' that Hayles (1995) discerns among disparate theoretical efforts to rupture the terms in which the question of nature has been posed. I take these emphases to imply an epistemological insistence on the situatedness of knowledge and a 'modest' ontological stance towards the performativity of social ordering.[5] At its most skeletal, 'hybrid geographies' takes a radical tack on social agency manoeuvring between two theoretical commitments. The first is to the de-centring of social agency, apprehending it as a 'precarious achievement' spun between social actors rather than a manifestation of unitary intent (Law, 1994: 101). The second is to its de-coupling from the subject/object binary such that the material and the social intertwine and interact in all manner of promiscuous combinations (Thrift, 1996: 24).

My aim in this book is to elaborate these stances not in the abstract but by working them through closely-textured journeys that follow various socio-material imbroglios as they are caught up in, and convene, the spatial practices of science, law and everyday life. It is organized as a series of paired essays that can be read in at least three ways: as cross-cutting conversations that interrogate the theoretical currents set in motion in this introductory chapter; as thematic sections that explore the spatio-temporal vernaculars of wild(er)ness, governance and consumption; or as individual essays that follow the interferences of 'things', from elephants and soybeans to deeds and patents, in the geographies of social life. This iterative style of argument works towards multiple mappings of the ethical import of taking hybridity seriously in/as geographical practice in terms of 'the real consequences, interventions, creative possibilities and responsibilities of intra-acting within the world' (Barad, 1999: 8). In the same spirit, these introductory orientations are not restricted to this chapter but continue to crop up as prefaces to each section, situating the particular essays they introduce.

My elaboration of these themes engages with four main bodies of work that converge through conversations between geography and science and technology studies,[6] but are also becoming aligned in more ambitious and various ways that Nigel Thrift has dubbed 'non-representational theory' (1999, 2000a, 2000b).[7] The first is science and technology studies (STS) where the vocabularies of hybridity have been most keenly honed through devices like the 'hybrid collectif' (Callon and Law, 1995), the 'quasi-object' (Serres and Latour, 1995) and the 'cyborg' (Haraway, 1985). 'Hybrid geographies' diagrams between the technical and corporeal emphases of two STS communities, those charily associated with the acronym ANT (Actant–Network Theory) (see Law and Hassard, 1999) and those of feminist science studies (see Haraway, 1997). My interrogations of these different efforts to accommodate 'non-humans' in the fabric of the social work to evince three consequences of such a redistribution of social agency. These involve shifts from intentional to affective modalities of association; from being to becoming in the temporal rhythms of human/non-human difference; and from geometries to topologies as the spatial register of distributed agency.[8] Above all, I want to hold on to the sense in this work in which 'the world kicks back' (Barad, 1998), or as Latour puts it, 'things' can object to their social enrolments (2000). At the same time, though, I want to exceed the scientific onus of these concerns and to mobilize the political implications of this 'redistribution' through other knowledge practices, notably those of law and governance, and everyday life.

The second engagement that situates this project is with bio-philosophy, which is never far from the various manoeuvres of science studies, particularly ANT (see Ansell-Pearson, 1999; Lorraine, 1999). Here, my argument is drawn into concerns with the morphogenic impulses of replication and differentiation, multiplicity and singularity through which the flux of worldly becomings takes, holds and changes shape. This rich vein of work folds debates on the philosophy of organism in the early twentieth century (such as Weisman, 1892; Bergson, 1983/1907; Whitehead, 1929) into those at its close, interrogating the precarious register of 'life' as a means of thinking past the human. Notable here are Deleuze and Guattari's vital topology (1988/1980), Bateson's ecology of mind (2000/1972) and Serres' material semiotics (1985).[9] 'Hybrid geographies' pursues this work's commitment to what Keith Ansell-Pearson calls the 'inherently ethical task of opening up the human experience to a field of alterity' (1999: 2).

The third theoretical conversation exercised in this book is with aspects of the diffuse literature on corporeality that have been particularly, though by no means exclusively, elaborated in feminist work (see Welton, 1999). Here, my argument engages with (various) theories of bodily *practice*. These serve both to reassert the corporeal affordances in which cognition inheres and, just as importantly, to challenge the cognitive privilege by extending the affective register of senses, feelings and habits engaged in 'thinking through the body' (see, for example, Radley, 1995; Weiss, 1999).[10] Haunting these debates is Merleau-Ponty's ontology of the flesh (*la chair*) and its emphasis on the reversibility of energies between bodies and worlds such that 'the touch is formed in the midst of the world and as it were in things' (1968: 134).[11] Taking the ethical import of this intercorporeal stance at its word, I interrogate these arguments by fleshing out the place of animal body-subjects in the geographies of wildlife conservation.

Last, but not least, 'hybrid geographies' engages work that is concerned with the knowledge practices of everyday life or what Shotter refers to as a 'third kind of knowledge' (1993). Here the notion of thinking through the body is invested in a particular direction, admitting the know-hows, tacit skills and bodily apprehensions through which everyday life goes on into the repertoire of knowledges that social/scientists need to take seriously (see de Certeau *et al.* 1998; Schatzki *et al.* 2000). These everyday knowledge practices have been argued to be performative rather than cognitive, such that 'talk' itself is better understood as action rather than as communication (see Shusterman, 2000; Thrift, 2000a). Allied to this argument is the suggestion that the spatialities of everyday life constitute a mode of dwelling, as against building, in the world (see Ingold, 1995a; Thrift, 1999). These arguments have a particular resonance for my determination to escape the scientific 'power-houses' of knowledge production and interrogate the ways in which nature–culture hybrids are apprehended through activities like consumption, and their interferences resisted and accommodated in the intimate fabric of social life (see Hansen, 2000a). These arguments are explored in the last section of the book in relation to the dissonance between consumer and producer knowledge practices in the event of food scares.

Thus freighted, the hybrid invites new ways of travelling that are beginning to make their mark in Geography (see, for example, Bingham, 1996; Murdoch, 1997a; Hinchliffe, 1999) and elsewhere (see, for example, Mol and Law, 1994; Strathern, 1996; Hetherington, 1997c). In place of the geometric habits that reiterate the world as a single grid-like surface open to the inscription of theoretical claims or uni-versal designs, hybrid mappings are necessarily topological, emphasizing the multiplicity of space-times generated in/by the movements and rhythms of heterogeneous association. The spatial vernacular of such geographies is fluid, not flat, unsettling the coordinates of distance and proximity; local and global; inside and outside. This is not to ignore the potent affects of territorializations of various kinds, just the reverse. It is a prerequisite for attending more closely to the labours of division that (re-)iterate their performance and the host of socio-material practices – such as property, sovereignty and identity – in which they inhere.

This book is not a lot of things. It does not espouse a particular philosophy, although its engagements and commitments position it philosophically. It is neither a complete 'thesis' nor an assembly of 'empirical' fragments, but rather an effort to germinate connections and openings that complicate this settlement. It is not a 'geography of nature' – though natures and geographies are always in play. Doubtless this list will grow as the book travels. Geography is at its most affective when, to use Homi

Bhabha's evocative phrase, the 'unhomely' stirs (1997: 445). In some sense, I owe my career as a Geographer to just such a fleeting fusion of the space-times of empire, discipline and self which occurred as I crossed the threshold between students and staff in the Geography Department at University College London.[12] That momentary slippage between worlds has shaped the kinds of geographical journeys I have sought to make ever since. But it has taken me more than a decade to venture a mode of geographical practice that holds on to this affect. It takes much of its inspiration from Game and Metcalfe's wonderful book *Passionate sociology* (1996) and its salutary immersion in life, compassionate involvement with the world and with others, and sensual and full-bodied approach to knowledge. Fleshing out a practice that shares these commitments but endeavours to enlarge the company of 'others' that they bring to notice has been a collaborative and heavily indebted activity, as is acknowledged in the preface. Not least in this company are the various 'companion-guides' (Bingham and Thrift, 2000) from Roman 'leopards' to Roundup Ready™ that I have enrolled in these journeys, as they have enrolled me in theirs. But it is also a question of style. Writing is an important part of any geographical practice (Barnes and Duncan, 1992). Indeed, as de Certeau suggests, stories *are* spatial practices that bear within them ghostly reminders of our journeying to and fro; they convey in words a sense of the body-subject occupying, inhabiting and traversing space, transforming it into places and specific presences (1988, see also Rogoff, 2000). In these essays I experiment with different ways of retaining the partiality and open-endedness of this 'to-ing and fro-ing' against the alliance of narrative and analytic conventions in social science that would forge it into completeness. 'Rather than vainly denying the living power of stories, an acknowledgement of narrative textures puts stories in their place' (Game and Metcalfe, 1996: 50).

As I hope is clear by now, the journeys undertaken here are not destined to arrive in the brave new world of a 'third nature' emerging perfectly formed from the 'machinic totality' of 'contemporary global capitalism' in which *everything* is caught up' (Luke, 1996: 11). In contrast to the universalizing ambitions of such accounts, the hybrid geographies that I work towards here are inescapably partial, provisional and incomplete. Refusing any vantage point that purports to take in the world at a glance, they are more modest in the claims they can, and want, to make and, by the same token, are more attendant to the energies of those they make claims about. Finally, such hybrid geographies work to invigorate the repertoire of practices and poetics that keep the promise of the Geographical craft alive to the creative presence of creatures and devices among us and the corporeal sensibilities of our diverse human being.

Section 1

> *Wild*ness (as opposed to wilderness) can be found anywhere; in the seemingly tame fields and woodlots of Massachusetts, in the cracks of a Manhattan sidewalk, even in the cells of our own bodies. (William Cronon, 1995: 89, original emphasis)

What does it mean to be 'wild' at the beginning of the twenty-first century? Everyday understandings of the 'wild' place the creatures and spaces so called outside the compass of human society. In various ways this treatment of the wild as a pristine exterior, the touchstone of an original nature, sets the parameters of contemporary environmental politics. Millennial anxieties about the seemingly limitless technological reach of human society, from global warming to genetic engineering, have shaken this framing of the wild to its core, a portent for some of 'the end of nature' (McKibben, 1989). Coming to terms with the contradictions of our own ubiquitous presence in the practices and spaces of wildlife management, tourism and multimedia, to name but a few, heralds important ethical and practical shifts in the life prospects and cultural freight of the creatures who inhabit this designation. Moreover, their import reverberates much closer to home. For at the very moment that the 'human mastery of nature' appears to have arrived, so the safety net that holds 'us' (humans) and 'them' (other animals) apart unravels as the instruments of this supposed mastery render our own species genome just one more entry in the vast informatic menagerie of life science (Cole, 1997).

The chapters in this section set out to explore the limits of these precarious geographies of wildlife, deterritorializing the creatures and spaces encapsulated by the wild to entertain more promiscuous patterns of worldly inhabitation that re-cognize its cargo of uncanny, but much less distant, kinds. Rather than an exterior world of original nature, I start with the premise that animals (and plants) designated wild have been, and continue to be, routinely caught up within multiple networks of human social life. These social orderings of animal life confound the moral geographies of wilderness which presuppose an easy co-incidence between the species and spaces of a pristine nature, confining their place to the margins and interstices of the social world. The chapters in this section trace a more volatile and relational conception of the topologies of wildlife that configure human and animal categories and lives in intimate, if not necessarily proximate, ways.

But, as the distinguished North American environmental historian William Cronon found in response to his remarkable essay 'The trouble with wilderness' (1995), these are dangerous waters indeed.[1] To question the sanctuary of wilderness is to disturb the orthodox parameters of environmental concern and to risk the wrath of those who, bolstered by scientific and/or environmentalist credentials, have cast themselves as custodians of the wild. Thus, for example, in an environmentalist slant on the so-called 'science wars', to entertain such questions has been condemned as intellectual 'tinkering' that is 'just as destructive to nature as bulldozers and chainsaws' (Soulé and Lease, 1995: xvi). In this climate, venturing into the wild – whether in the scientific guise of the biodiversity reserve or the environmentalist guise of the sacred grove – is unavoidably bound up with passions and convictions that enmesh personal, political and professional sensibilities in potent and complex ways, including my own.

In his essay, Cronon lays bare the historical erasure of 'indigenous' peoples, both figuratively and physically, which underwrites the wilderness premise that nature, to be natural, must also be pristine. The uncomfortable burden of his argument is directed at the political discourses of (North American) environmentalism rooted in this purification of the spaces of 'nature' and 'society' (Haila, 1997).[2] These discourses span the measured tones of established conservation bodies like the Wilderness Society and Sierra Club, which combine the vocabularies of nineteenth-century nature romanticism and contemporary conservation science, and the militaristic rhetoric of a new breed of 'eco-warriors' whose stated mission is the defence of 'the big outside' (Foreman, 1981). Such discourses, Cronon argues, 'get us back to the wrong nature' (1995: 69) in the sense that they reproduce categorical binaries between society and nature, human and animal, domesticated and wild that are intellectually and politically moribund.

Playing on Thoreau's famous dictum, the opening quotation from Cronon's essay signals the importance of geographical imaginations and practices both to keeping 'nature' and 'society' in their proper place and to freeing them from this binary fix. Given the discipline's instrumental role in mapping the 'wildernesses' of European colonization (for example, Driver, 1992; Livingstone, 1992), and the currency of profoundly geographical concepts like landscape and ecology in the accounts of other disciplines today, geographers have paid remarkably little attention to wildlife (Philo, 1995).[3] Only now, with significant moves to reverse the neglect of animal life in the social sciences (for example, Arluke and

Sanders, 1996; Ham and Senior, 1997; Wise, 2000), are alternative
geographies beginning to emerge that admit more agents than humans
and other spaces of action than 'outside' (see Wolch and Emel, 1998;
Philo and Wibert, 2000).

Taking a leaf out of Elspeth Probyn's book *Outside belongings*
(1996), I want to flesh out this topological conception of wildlife through
close textured examples and ways of narrating them which disconcert the
space–time coordinates of the wild – the syntax of distance and
proximity; inside and outside; then and now – by juxtaposing historically
and geographically remote configurations of wild-life. In chapter 2 these
arguments are worked through glimpses of two historically very different
social orderings of 'wild' animals – those associated with the military
vernacular of the gladiatorial games of Imperial Rome and the scientific
vernacular of endangered species listing and conservation under CITES
(the Convention on the International Trade in Endangered Species) today.
These 'foldings' of wildlife in distant time–spaces aim to disrupt the
linear historical narratives of 'civilization' and 'evolution' which consign
wildlife to marginal spaces that share a teleological destiny of erasure.
Chapter 3 takes up the most intimate of these 'foldings' to explore the
ways in which the embodied experiences of particular animal kinds are
performed in and through such configurations of wildlife. Here I trace
multiple moments and spaces incorporated in the tricky and thoroughly
situated business of becoming elephant, in this case 'the' African elephant
(or *Loxodonta africana* in zoological taxonomy), in two contemporary
networks of wildlife conservation. The first is concerned with the
computerized management of animals held in zoological collections
worldwide for the captive breeding of so-called *ex-situ* wildlife
populations, while the second recruits paying volunteers to scientific
expeditions investigating wildlife populations *in-situ*. These visceral
tracings of animal lives aim to unsettle their taken-for-granted status as
material objects and consider the theoretical consequences of admitting
them as radically different kinds of subject into the company of the
social.

Displacing the Wild:
topologies of wildlife

> What she had seen from that building at Aldgate was a city that stretched to the ends of the earth . . . Madelene saw that . . . any zoo, any game reserve, any safari park . . . was now contained within the bounds of civilization. . . . She turned to face the ape. 'There's no such thing as outside now,' she said. 'If there's any freedom to be found it'll have to be on the inside'. (Peter Hoeg, 1996: 74)

heterotopic si(gh)tings

The wild occupies a special place in the imagined empires of human civilization as that which lies outside its historical and geographical reach, however defined (White, 1978). A place without *us* populated by creatures (including, surreptitiously, a variety of human 'kinds') at once monstrous and wonderful, whose very strangeness gives shape to whatever *we* are claimed to be. The enduring coincidence between the species and spaces of wildlife as the antipodes of human society means that to ask what is wild is always simultaneously a question of its whereabouts. This framing of the wild renders the creatures that live 'there' inanimate figures in unpeopled landscapes, removing humans to the 'here' of a society from which all trace of animality has been expunged (Macauley, 1997). As Madelene Burden, the heroine of Peter Hoeg's novel about a love affair between a woman and an ape, comes to realize during her rooftop flight across London, to question what it means to be wild is to disconcert this binary geographical imagination and entertain forbidden possibilities for being otherwise in the world.

Hoeg's novel resonates with what has become something of a truism among environmental historians, that this placing of the wild as a pristine exterior harbours within it the very will to power that the gesture would elude (see Buell, 1995). While environmental historians, notably in North

America, have done much to expose the workings of a particular (linear) historicity in the wildernesses wrought by the impulses of modern European colonialism (Grusin, 1998), rather less attention has been paid to the workings and consequences of the particular geo-graphies that configure these otherworldly space–times. Some of the most provocative work in this direction has sought to frame wilderness in terms of Foucault's interrogation of utopia, as a construct of exteriority characteristic of a peculiarly Modern spatial imaginary (see, particularly, Birch, 1990; Chaloupka and MacGreggor Cawley, 1993; Macauley, 1997). Utopias are imaginary spaces in which the abstracted essence of what society is not can take shape as an 'outside', and thereby provide vantage points for social critique (Hetherington, 1997a). Figured thus, wilderness stands as the transcendent sign and site of the radical otherness of a nature without a past; an immaculate space defiled by any taint of human presence. This is the political terrain of environmental direct activists like EarthFirst!, for whom

> [t]he only hope for Earth (and humanity for that matter) is to withdraw huge areas as inviolate natural sanctuaries from the depredations of modern industry and technology . . . that can be restored to a semblance of natural conditions, . . . and declare them off limits to modern civilization. (Foreman, 1981: 41)

Foucault's notion of heterotopia, by contrast, forces us to confront these volatile exteriorizations as places of our own making, configured in relation to the interiorized sites of knowledge, imagination and desire (Foucault, 1973, 1986). It has been highly influential in human geography as a means of exploring carceral spaces – the poorhouse; the asylum; the prison – in which the 'outside(r)s' of various orderings of social life take shape as counter-sites in the fabric of the modern city (see, for example, Driver, 1985; Gregory, 1994; Soja, 1996). Refigured in these terms wilderness is inextricably social and becomes a disturbing and disruptive place in which, as Chaloupka and McGreggor Cawley put it, the 'open secret' of a dense and energetic infrastructure of wilderness management can be exposed (1993: 11). Here, as the protagonists of Peter Hoeg's novel urge, the futures of earth creatures (including humans) would seem to lie not in fortifying the utopian space–time of a pristine wilderness, but on the inside where the everyday worlds of people, plants and animals are always already in the process of being mixed up.

I want to explore this line of argument further but with the proviso that such a heterotopic re-cognition of the wild does not, as the opening quote suggests, mark some kind of radical break between human–animal relations in the late twentieth century and those that have gone before. While Foucault attached particular significance to Modern heterotopia, he considered them to be a widely established feature of the spatial repertoires of human societies, from sacred sites and forbidden places to theatres and gardens. In similar vein the contention here is that 'wild' animals have been, and continue to be, routinely imagined and organized within multiple social orderings in different times and places. Their myriad (re)positionings within these networks have been complicating animal geographies long before the possibilities of genetic engineering startled our commonsense coordinates of the place of the wild. This is not to ignore the historical specificity and cultural potency of wilderness as a utopian space in the Modern spatial imaginary, but to resist the moral vortex that it has come to represent in conservation discourses and practices.

Efforts to refigure wilderness as a heterotopic space, on the 'inside', are an important first step to challenging the binary geographies of 'nature' and 'society' and the associated purifications of human and animal lives. However, they do not go far enough to advance the kinds of hybrid geographies of wildlife that I am working towards here. In particular, they retain an exclusively human framing of the social fabric of such spaces and privilege the optical over other sensory registers in rendering them affective (see, particularly, Chaloupka and McGreggor Cawley, 1993). This has (at least) two undesirable consequences for my purposes. First, it evacuates the bodily presence of living creatures from the si(gh)ting of the wild. The corporeal spaces configured in the process of becoming animal (and human) are, thus, removed from the compass of analytical consideration. Secondly, it erases all but 'humans' as agents in the making of these wild places. The diverse energies of all other earthly inhabitants (and the earth itself) get rolled into a lumpen 'nature' that amounts to little more than 'a substrate for the external imposition of arbitrary cultural form – a tabula rasa for the inscription of human history' (Ingold, 1993: 37).

In contrast, the notion of wildlife being fleshed out here is a relational achievement spun between people and animals, plants and soils, documents and devices in heterogeneous social networks which are performed in and through multiple places and fluid ecologies. Two manoeuvres take us from the familiar utopian spaces of pristine nature as wilderness to these more promiscuous topologies of wildlife. The first is to unsettle the contours of these exteriorisations of the wild by situating them within the diverse currents and flows through which multi-sited wildlife networks are config- ured. Here, 'a thing's place [is] no longer anything but a point in its movement . . . [a] space that takes for us the form of relations among sites' (Probyn, 1996: 11, after Foucault). The second manoeuvre involves ani- mating the creatures mobilized in these networks as active subjects in the geographies they help to fashion. Their constitutive vitality is acknowl- edged not in terms of unitary biological essences but as a confluence of libidinal and contextual forces. Here, the multi-sensual business of becom- ing, say, antelope or wolf and the inscription of these bodily habits in the categorical and practical orderings of human societies are interwoven in the seamless performance of wild-life.

These two moves towards a performative conception of wildlife, as a relational and fluid achievement, render the experience of radical difference delineating the human from the animal, the civilized from the wild, as a con-figuring – a drawing together, as Jennifer Ham puts it – rather than a holding apart (1997). In so doing, they begin to open up new possibilities for addressing the pressing dilemmas engendered by these boundaries. Most significantly, the moral high ground starts to shift from an unerring pre-occupation with shoring such boundaries up to the painstaking business of tracing the historical and geographical particularity of human–animal relations as a condition of securing their ongoing dynamism and diversity. I return to consider the ethical dimensions of such a shift later in the chapter but, first, want to work through some of the implications of this proposed refiguring of wildlife in terms of networks and bodies more closely, by example.

I begin by describing the topology of devices, documents, bodies and sites through which two historically remote wildlife networks provisionally take, and hold, their shape. These are the networks bringing wild animal participants to the *venationes* staged in the Roman games and to the management of 'endangered species' under the Convention on the International Trade in Endangered Species (CITES) today. I then journey through them in the company of two of their animal inhabitants, the leopard and the broadnosed crocodile. More precisely, these creatures are accompanied in their particular manifestations as *leopardus*, the Roman term referring to a number of 'spotted cats' chiefly of north African origin, and *Caiman latirostris* in the taxonomic nomenclature of modern science.

topologies of wildlife

> The dividing line between nations may well be invisible; but it is no less real. How does one cross that line to travel in the nation of animals? Having travelled in their nation, where lies your allegiance? What do you become? (Montgomery, 1991: 209).

In very different ways, through military and scientific modes of ordering respectively, the Roman games and the Species inventories of our own times have been widely understood as imperia – as the spreading outwards of some unitary and irresistible force across the surface of the known/civilized world. Paralleling the heterotopic reading of wilderness, the amphitheatre arena and the nature reserve might readily be identified as the most potent sites for the performance of wildlife in these contexts. However, I want to render the power of these modes of ordering in more relational and distributed terms, spatially configured through currents as much as sites;

bodies as much as places. To do so I begin by situating the fabric of the amphitheatre in which the bloody spectacle of the Roman games was staged in wider networks of people, instruments, documents and places assembling these wildlife performances through the capture, transport and training of animals. In the case of the Species inventories of today, the classificatory practices of Science inscribed in the bodies of animals themselves are pivotal to the measure and reach of global conservation conventions. Here, I start by situating the inscription of *Caiman latirostris* in wider networks of people, instruments, documents and places, assembling these wildlife performances through the monitoring, management and (de)listing of animals as endangered species.

wildlife networks

Roman games i

The site and spectacle of public games played a key part in Roman life, nourishing imperial power and reach through staged demonstrations of the compass of Rome.[1] Peoples and creatures from its furthest outposts were pitted against one another to the death for the entertainment of those whose presence in the ranks of spectators affirmed their place in the Roman body politic (Plass, 1995; Futrell, 1997). The monumental architecture of amphitheatres and circus arenas, in which these games were staged in towns throughout the Roman empire, remains impressive to this day (see figure 2.1). The largest, the Colosseum in Rome, completed in AD 80 under the Emperor Titus, held some 50,000 people and reputedly witnessed the killing of some 9,000 wild animals (*ferae*) during its inaugural festivities. What began as a calendar of religious events became over time more secularized, extravagant and bloody affairs, instigated by prominent military and political figures to mark their own achievements or to satisfy the expectations of the people in whose name they ruled. Lasting anything from five to twenty days, these popular spectacles became keenly political performances of the rights and hierarchies of Roman civic society (Auguet, 1972).

Games took place in several forms; most common were *ludi circenses*, circuses in which the main event was chariot racing but which included many forms of trained animal acts. Most extravagant were the staged hunts in which arenas were designed to resemble forests (*silvae*), or even flooded for 'sea fights' (*naumachiae*). But those which excited the greatest passions and largest followings were the *munera* (gladiatorial combats) and *venationes* (wild animal combats) which came to involve all manner of incarnations (human and animal) of the strange and uncivilized regions of the Roman imagination. These games took place on ever grander scales over a period of some 900 years between the founding of the Roman

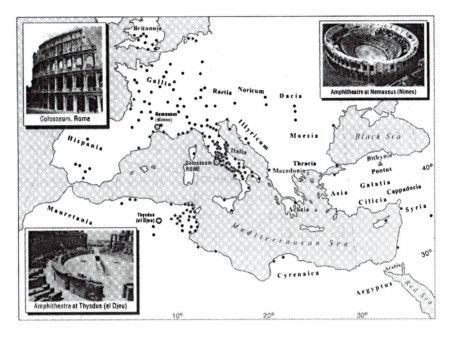

Figure 2.1 Siting the Roman games (second century AD) (adapted from Scarre, 1995: 83)

Republic in 510 BC and the banning of gladiatorial combats by the Emperor Honorius at the beginning of the fifth century AD.

The *venationes* became highly choreographed performances of the wild, involving ever more intensive training of human and animal combatants alike. They took a wide variety of forms, ranging from 'wild' animals pitted against each other or specialist animal fighters (*bestiarii*), to their use as instruments of execution in the dispatch of people condemned to death by the Roman state (including criminals, deserters and escaped slaves, enemy captives and persecuted religious minorities) (Jennison, 1937). *Bestiarii*, for the most part, were drawn from these same outcast social ranks and denied the rights of Roman citizenship. As such, they were seen by their contemporaries as closer in status to the beasts they fought than to the civilized spectator. Leading Roman commentators may have found popular tastes unedifying but they endorsed the martial morality of the *venationes* as an opportunity for their human and animal participants, unlike those condemned to death, to redeem themselves through combat and so die honourably (Wiedemann, 1992). Such events figured prominently in the material culture of the period, or at least in those documents and artefacts which have found their way into the modern archaeological record and scholarly accounts of the Roman world (Toynbee, 1973; Ville, 1981). Some sense of the variety of these stagings of the wild can be gained from images in mosaic decorations surviving from public and private

Roman buildings, like those from a villa floor in Nennig shown in figure 2.2.

The spectacular carnage taking place in Roman amphitheatres was the focal point and public face of elaborate networks of people and animals mobilized by military conquest, political patronage, administrative taxation, legal or judicial ruling and commercial trade. As the imperial network of Rome expanded through successive military campaigns, so the spectacle of the games as both demonstrations of the power of their patrons and the tastes of their audiences became bound up with the exotic. At its height in AD 117, under the Emperor Trajan, the threads of empire stretched the presence of Rome throughout the Mediterranean and Adriatic region to the Atlantic in the west (including Britain); the coastal regions of North Africa to the south (including Egypt), Assyria and Armenia to the east and as far as the Rhine and the Danube in the north (Scarre, 1995). Enormous efforts and monies were invested by Roman patrons in procuring unusual and prized species of 'wild' animal for their games over considerable distances.

Figure 2.2 Mosaic reliefs immortalizing *Africanae bestiae* in combat (Auguet, 1972: plate 17)

Animals were recruited through two main networks of procurement. First were the imperial conduits of military supply lines and political patronage which connected the distant lives and spaces of the Roman Senate with those in the Provinces. Second were the commercial trading routes by land and sea which stretched well beyond the shifting borders of the Roman empire, particularly to China and India to the east. The most important of these commercial networks developed as Associations, specializing in all aspects of animal supply to menageries, stud-farms and the exotic pet market, as well as to gladiatorial schools (Charlesworth, 1924).

Jennison, in his classic account of animals in ancient Rome (1937), gives an example from the correspondence of Cicero of such a net-working. Cicero was appointed governor of Cilicia in southern Asia Minor in 51 BC. Hardly had he arrived there, than he found himself the recipient of ever more desperate epistles from his friend Marcus Caelius Rufus requesting leopards for games he was preparing in anticipation of his being elected to the *curule aedileship* in 50 BC.

> In nearly all my letters to you I have mentioned the subject of leopards. If you will only remember to set the Cibyrates to work . . . – you will get what you want done. . . . Do please see that you attend to this. . . . In this affair the trouble for you is only to talk – I mean, to issue orders officially and to give commissions (*imperandi et mandandi*). For as soon as the leopards are caught, you have my people, whom I have sent [on another matter] to look after the animals' keep and bring them to Rome. I think, too, if you write encouragingly, I shall send some more men of mine to your part of the world. (Jennison, 1937: 137–8, original translation)

Cicero puts off any response until Caelius's *aedileship* is certain and even then, reluctant to press local elites with official orders for a municipal hunt so early in his appointment, he merely commissions some commercial hunters to capture leopards. In his letters to Caelius he also reports the scarcity of *pantheras Graecas*, a creature which Roman appetites seem to have exhausted altogether in common with several other animal extinctions (Hughes, 1994). We do not know if Caelius ever received any leopards, but instigating a hunt would have been only the first of several precarious actions at a distance that would have been necessary to bring the wild to life in a Roman *venatio*. I will return to some of the other threads in this wildlife network below, in a second glimpse of *leopardus* on the move.

Caiman latirostris i

Living in the swampy reaches of certain South American water ways, a broad-snouted crocodilian entered the networks of science almost 200 years ago. The folk names for this kind of animal included *yacare overo*, or *ururau*, or *yacare de Hocico Ancho*, depending on its location. But in 1801 the French explorer and zoologist Daudin situated the crocodilian within a scientific taxonomy that was becoming standard, by naming it *Caiman latirostris* (see figure 2.3) (Spary, 2000). Thus enlisted into the project of science, these animals became instrumental in extending the compass of its knowledge claims through sightings, dissections and mappings carried out in a variety of material forms. In the process, the viability of their status as wild animals has become the subject of scientific expertise which, latterly, has itself become intertwined with the practices of wildlife trading.

The tenth edition of the Swedish naturalist Carl Linnaeus's *Systema naturae*, published in 1758, is taken as the edition that 'forms the starting point for all generic and specific names of animals, previous names being ignored' (Freeman, 1972: 2). The name of a genus anchored the classificatory system, while the species name was chosen in any of a number of ways (with reference to place, colour, body marking or shape, for example). Together these two words constituted the scientific name, always printed in italics, hence, *Caiman latirostris*. Through this nomenclature, an animal obtains a fixed and unique identity which marks its position in the (known) animal kingdom (Ritvo, 1997). The Linnaean Society established in 1788 promulgated this system as a universal standard for the expanding scientific community associated with European colonialism (Koerner, 1996). Despite significant shifts in the onus and practice of natural history, this binomial classificatory system entered the rules of nomenclature established by the

Figure 2.3 Pixelled web image of *Caiman latirostris* 'in the wild' (www.flmnh.ufl.edu/natsci/ herpetology/crocs/crocsb.htm)

International Congress of Zoology (and Botany) at the end of the nine-teenth century and underpin the biological sciences to this day (Systematic Zoology, 1959).[2]

The process of scientific naming classifies animals by means of a species identification, principally through morphological comparison, dis-tinguishing and fixing its relationship to all other animals. The year after his description of *Caiman latirostris*, Daudin identified a species of similar appearance and with overlapping range as biologically distinct and named it *Caiman yacare*. In the process of drawing up such inventories, animals were removed from their environmental and social context and preserved as unique specimens by diagrammatic or corporeal means. Through their depiction as organic machines, disassembled and mapped anatomically (as heads, transections, skeletons, embryos, etc.) in zoological illustrations, animals became mobilized as species through the expanding networks of science (Jardine *et al.*, 1996). The physical translocation of specimens (dead and alive) followed, establishing biological storehouses of exotica in mu-seums, menageries and amateur collections in the heartlands and outposts of European empires (Sheets-Pyenson, 1988). This colonial impulse to subject the world to systematic scientific account proved itself to be 'one of the most authoritative actions in the exercise of government' (Raby, 1996: 5). Entangled in these networks, the bodies and places of animal inhabita-tion became standardized, portable facts, their 'species distribution' a matter of expert determination from afar.

By the twentieth century, animals (including humans) found them-selves the subject of ever more intensive biological scrutiny as expanding numbers of career scientists and investigative devices fleshed out the minutiae of their bodies, cells and genes.[3] The latest, and most far-reaching, enactment of this empire of knowledge is the newly minted science of biological diversity – a 'new definition of nature' being vigorously pro-moted, according to David Takacs, by an influential 'cadre of ecologists and conservation biologists' (1996: 1). Biodiversity, as it is colloquially termed, refers in one and the same breath to

> the richness and variety of life on Earth. The flowers and insects and
> bacteria and forests and coral reefs are biodiversity. . . . [and to] an area
> of scientific research, including both description and measures of diver-
> sity and explanations of how this diversity is created. (Jeffries, 1997: 3)

The scientific networks of biodiversity include the fields of systematics, ecology, population biology, animal behaviour and comparative biology (Cracraft, 1995). The first of these, systematics, fuels the scientific zeal for classification with a new moral purpose. This is set out most clearly in the US-inspired Systematics Agenda 2000 (subsequently endorsed by the Lin-naean Society), whose objectives include the mapping of global species diversity over the next 25 years. Here, the project of classification is no longer seen 'just' to contribute to our knowledge of biological complexity, but 'directly impinges on the integrity of biological knowledge and the values we place on biodiversity' (Cotterill, 1995: 183).

These knowledge practices and values are becoming increasingly institutionalized through their codification in the regulatory networks of global environmental management. Most significantly, their emphasis on systematics and taxonomy has become the watchword of the newest star in the constellation of international environmental treaties, the Convention on Biological Diversity (CBD). The Convention's objectives are: 'The conservation of biological diversity, the sustainable use of its components and the fair and equitable sharing of the benefits arising out of the utilization of genetic resources' (CBD, 1992: Article 1). The logic here is to establish a universal exchange rate between the scientific value of animal (and plant) life, measured in units of biological rarity, and that most pervasive of human currencies – economic value. Its reach is being extended to other networks of global environmental management by means of Memoranda of Cooperation. For example, the Memorandum signed between representatives of the CBD secretariat (based in Montreal) and the CITES secretariat (based in Geneva) in Brisbane in 1996 urged the nations party to these Conventions to promote 'effective conservation and . . . the sustainability of any use of wildlife as a part of the biological diversity of our planet' (1997a: Article 4).

Thus, the global science of biodiversity has formalized an ongoing relationship with the institutions responsible for regulating the international wildlife trade. Among the organizations collaborating in the practice of 'biodiversity management' are the Species Survival Commission (SSC) of the International Union for the Conservation of Nature (IUCN), which is 'organized primarily along taxonomic lines' (Rabb and Sullivan, 1995). Disseminating information about species and their status worldwide is the remit of another organization, the World Conservation Monitoring Centre (WCMC), which hosts the CITES secretariat website. These organizations are daily engaged in coordinating a performance of wildlife as a biological resource which is spun between the codes and devices of scientific authority, living animals inscribed as specimens of more or less 'abundant' biological populations, and regulatory conventions on the trade in their body parts (Thorne, 1998). In the second glimpse of this wildlife network, below, I consider the ways in which *Caiman latirostris* is animated in the performance of an 'endangered' wildlife species.

animating wild-life

Thus far, I have sought to render the contours of key sites in these very different performances of wildlife – the amphitheatre arenas in which the

Roman *venationes* were staged and the animal bodies in which Scientific classifications are inscribed – as relational achievements, configured within heterogeneous networks and fluid topologies. Animal participants in the *venationes*, like the human *bestiarii*, were trained and coerced to play their part in the choreographed acts of violence and aggression that conjured the wild in the Roman imagination. Animal participants in zoological inventories today are no less disciplined by their naming to behave and reproduce themselves as living kinds that designate the wild in the scientific imaginary of global conservation. Perversely, the multi-sensual animality of the creatures caught up in these performances of wildlife all but disappears in both cases as they become symbolic and material units in some human currency – blood and death perhaps for the Roman crowd; genes and resources for the visionaries of planetary management. In the second of the manoeuvres outlined above, we catch another glimpse of these wildlife networks as if their animal participants mattered both as active agents and experiential subjects.

Roman games ii

One of the most popular and sought after animal participants in the Roman *venationes* were the large spotted cats known in contemporary parlance variously as *leopardus, panthera, pardus* and *varia* and recognizable in our own time by their species names *Felis pardus* (leopard) and *Cynaelurus jubatus* (cheetah). As Jennison explains, Roman zoological nomenclature, itself building on the Greek, was far from systematic both because of the uncertain identity of unfamiliar animals and the widely-held belief in the 'breeding of hybrid forms in nature' (1937: 183). The elder Pliny, in his encyclopedic *Natural history* (1991), for example, treats leopard (*varia* or *pardus*) and cheetah (*panthera*) as the male and female of the same species (VII, 17 (23)), while *leopardus* refers to the maneless offspring of the female lion (*leone*) and the male leopard (*pardus*) (VII, 16 (17)).[4]

Leopardi, in all their guises, were principally sourced from Africa. In Strabo's *Geography*, Africa is described as the third and 'very much inferior' portion of the habitable world (after Europe and Asia), due to its being 'for a great part . . . desert' and 'the nursery of wild beasts' (1916: 275). Roman Africa referred in more practical terms to the northern coastal strip of the continent running from Tangier in the west to Cyrene in the east (modern-day Morocco to Libya) which became a wealthy Roman province after the end of the Third Punic War in 146 BC (Manton, 1988). Here, leopards had long been caught up in the lives of local people whether as everyday threats to the domestic animals of nomadic and agricultural communities, prized goods in the trading caravans of the peoples of the interior or, like Hannibal's legendary elephants, in the pomp and ceremony of war and governance. Their mobilization as African beasts (*Africanae bestiae*) in the networks of Roman commerce and patronage harnessed these already relational performances of wildlife to the project of Romanization through the site and spectacle of the *venationes* (Auguet, 1972).[5] Journeying with the leopard through these networks, many hands, devices and places leave their mark on the creatures becoming *leopardus* (see figure 2.4).

Scenes and techniques of animal capture and transportation, like those of the *venationes*, are widely depicted in the decorative fabric of Roman buildings, most famously in the mosaic pavement of the 'Great Hunt' at a Roman villa near Piazza Armerina in Sicily, but not least in the cities of Roman Africa itself. These imaginative works record contemporary awareness of the processes involved in bringing animals to the arena and their significance in the aesthetics of public and private life.[6] They show Africans as well as Romans deployed in the hunt and using a range of techniques for capturing leopards, including the pit, the net and the trap. *Leopardus* would be drawn into a barricaded pit by the noise of a decoy animal fastened to a central pillar. A cage, baited with meat, would then be lowered into the pit and the leopard hauled up for transportation. Nets were used to effect the larger-scale capture of animals. In a picture dating from around AD 300 in a Roman villa in Bona (Algeria), an assortment of spotted cats is shown being driven by mounted Numidean beaters into a line of netting, disguised by foliage, which encircles them as a ring of Roman hunters on foot, wearing protective clothing and bearing torches, closes ranks (Jennison, 1937: 145). Alternatively, the wooden travelling-boxes used to transport a captive leopard to the nearest town or port could itself be used as a trap by luring the animal into it with bait or, in one depiction, by the use of mirrors (Toynbee, 1973: 83).

The captive leopard then faces a hot, thirsty journey by mule-drawn cart across unmetalled terrain before continuing on the Roman roadways linking agricultural regions and towns to the major cities and coastal ports like Carthage and Lepcis Magna. Confined within the metal bars of a cage (*cavea*) or a reinforced wooden crate (*claustra*), this arduous journey would be drawn out over a period of weeks by regular, sometimes scheduled, stops in outlying communities and watering places to secure food for the animals and their escort. A leopard making it this far alive might have ended its journey in Roman Africa in one of the amphitheatres or animal troupes that entertained military garrisons, administrators and the local populace (see figure 2.1 above). But most, after a shorter or longer sojourn on the harbourside, depending on the weather, would have found themselves on the move again, this time in a ship's hold. Wooden merchant vessels, like the *Europa* depicted in a plaster engraving on the wall of a house in Pompeii (Greene, 1986: 27), were busy in the movement of all manner of

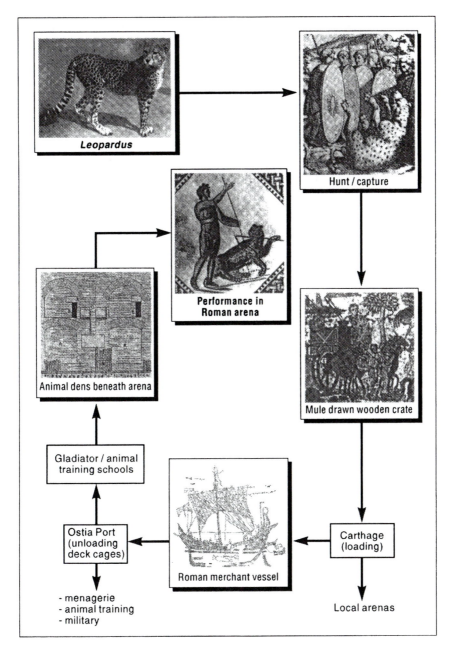

Figure 2.4 Becoming *leopardus* in the networks of the *venatio*

goods – grain, slaves, minerals, wine – around the Roman Mediterranean. Equipped with the triangular lateen sail, the ability of these vessels to tack and negotiate the winds brought the voyage from Carthage to Ostia or

Civitavecchia (outside Rome) to an average of 3–5 days.[7] But with navigation dependent on good visibility, the sailing season was largely restricted
to the period from mid-March to mid-November (Greene, 1986).

Deprived for a period of months of the sensory experiences, daily
habits and social bonds through which the leopard had made its place in
the world, and subject to the unfamiliar and gruelling conditions of its
journey, already the creature surviving its passage to Rome would have
come a long way in its refashioning as *leopardus*. Once the customs duty
(*portorium ferarum*) had been paid in recognition of their arrival, the
destination of most *leopardi* would have been a cage in the *vivaria* (animal
depots) attached to amphitheatres and gladiatorial schools. Several are
recorded on the outskirts of the city of Rome, a location designed to
minimize the public danger effected by the not infrequent escape of captive
animals (Jennison, 1937: 174–5). Here, *leopardi* would have been brought
into the hands of those most closely concerned with the disciplines of the
venatio. With their human counterparts, the creatures' skills and instincts
were honed to the moves and props of combat by the judicious use of
whips, torches and starvation. Those failing to perform by being 'inappro-
priately' ferocious, mauling a keeper or trainer say, or by failing to be
ferocious enough in combat were likely to meet the same end.

The night before a *venatio*, a starved, abused and often diseased
incarnation of a leopard from somewhere in Africa, would be taken to the
underground cages of the Colosseum along with hundreds of other animals. What did *leopardus* sense in this dark, rank place? The blood and
faeces of other creatures; a distant din of violence above; a wordless fear or
mindless rage? Raised in cages by windlasses, or carried in crates along the
inner corridor, to the place of entry to the arena, *leopardus* is driven into
the blinding sunlight and roar of the crowd by attendants brandishing
burning straw to its first, and usually last, wildlife performance. But even in
this final moment as its lifeblood ebbs into the sand, *leopardus* lives on in
vivid fragments of mosaic and painted plaster, like 'Crispinus' the *venatio*
star from Smirat still pleasing the crowds in the Sousse Museum (Manton,
1988: 108).

Caiman latirostris ii

As we have seen, *Caiman latirostris* has been subject to the curious
disciplines of natural history and science for some 200 years. More
recently, since Argentina became a party to CITES in 1980, the broadnosed crocodile has been listed under *Appendix I* as a scientifically
designated 'endangered species', thereby prohibiting trade. However,
among the swathes of paper that conference delegates carried to the tenth
meeting of CITES (COP 10) in Harare, Zimbabwe, in June 1997, Proposal

10.1 called for the down-listing of the Argentine population of *Caiman latirostris* to *Appendix II* under an established Resolution to admit ranching as an acceptable form of wildlife management under CITES (see figure 2.5).

As Lyster explains:

> A ranching operation is not closed-cycle like captive breeding but involves the rearing of wildlife, usually from wild caught eggs or young, in a controlled environment. Since they do not qualify for the captive bred exemption, specimens of *Appendix I* species cannot be ranched and then traded internationally for commercial purposes without violating CITES. At the San Jose Conference, however, delegates from several Parties argued that they could only justify protecting habitats of endangered species from agricultural and industrial development if they could derive some economic benefit from the species. (1985: 261)

Proposals for the ranching of crocodilians are compiled and reviewed by experts of the Crocodile Specialist Group (CSG) of the IUCN (see above) in cooperation with national CITES authorities. Funded from private donations, the CSG describes itself as 'a world-wide network of biologists, wildlife managers, government officials, independent researchers, non-governmental organization representatives, farmers, traders, tanners, fashion leaders, and private companies' (Crocodile Specialist Group, 1997a). From the outset then, the expertise mustered in the name of the crocodile is an eclectic mix of scientific, conservation, commercial and policy interests and rationales. Unsuprisingly, perhaps, the CSG has taken a pragmatic line on the conservation priorities for crocodilians, advocating what it calls a 'creative' approach whereby the (regulated) sale of crocodile bodyparts provides incentives to local people to ensure the species' survival. This is an example of the idea of 'sustainable use' which is gaining currency in global wildlife management networks more widely. As the CSG explain, 'sustainable use' is defined by the IUCN as 'an activity that can be continued indefinitely' (McNeely *et al.*, 1990). In practice, it is determined by expert assessment of

> . . . the effects on the target population (e.g. *Caiman* that we wish to hunt) and the effects on non-target species and the associated ecosystem (e.g. hunting *Caiman* may affect wetland nutrient cycles and fish populations). In many cases, it is difficult to define or prove that use is sustainable, but it is relatively easy to recognise when use is not sustainable. If people use any resource at a rate that exceeds the ability of the resource to replace itself, then the resource will become depleted and no longer be available for use. (Crocodile Specialist Group, 1997b)

Proposal 10.1 before delegates in Harare was, then, politically well-crafted, building on a succession of ranching resolutions over the previous

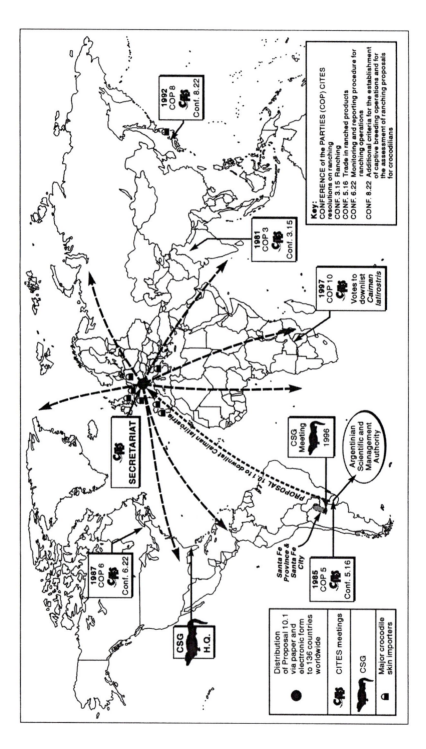

Figure 2.5 Organizational relations in global wildlife management

16 years and on the regional and scientific credentials of the chairman of the Latin American and Caribbean CSG. The conservation of *Caiman latirostris* was set to join several of its reptilian relatives as a hostage to the commercial value of its skin (Jenkins and Broad, 1994; Thorbjarnarson, 1999). Significantly, the proposal was endorsed by the CITES secretariat, whose staff confirmed that 'qualified scientific and technical staff' were working on the project. With an assurance about establishing a tight skin-marking scheme to enforce the distinction between ranched and 'wild' specimens, the conference passed the proposal for down-listing with little controversy.

The CITES secretariat and related agencies in the networks of global wildlife regulation are at pains to originate their decisions to list and delist species with the independent advice of expert scientists. The fate of *Caiman latirostris* in these global networks saw this account being challenged by other parties engaged in the politics of crocodilean conservation. A report by Traffic International, for example, questioned whether the reptile skin industry provided a very auspicious model for wildlife use, given '[an] almost complete lack of demonstrable sustainability and the absence of any significant linkage between the trade and conservation action at habitat or species level' (Jenkins and Broad, 1994: 63). The CSG Action Plan for *Caiman latirostris* itself admits that the problems of habitat destruction by farming experienced in Brazil have not encroached on the 'original habitats of the species' in Argentina. In this context, the human pressures requiring the 'creative' crocodile ranching incentives of Proposal 10.1 appear some-what conjectural. Likewise, the expert understanding of population dynamics identified by CSG as a prerequisite for 'sustainable use' programmes, does not seem to be entirely consistent with scientific knowledge about *Caiman latirostris*. The IUCN Red list for 1994 estimates the population to be 10,000 and widespread. Yet survey data for the species is described as 'poor', due largely to habitat inaccessibility. This is confirmed by CSG's own report which notes that 'due to a lack of field studies, little is known about the behaviour and ecology of this species. Much of what is known about reproduction has come from individuals in captivity' (Crocodile Specialist Group, 1997b).

The downlisting of *Caiman latirostris* permits Argentina to trade the skins of 12,500 ranched animals in the period 1998–2000, within the terms of the CITES Convention. The ranching process (see figure 2.6) begins with the collection of eggs by appointed scientists from 'wild' crocodile nests within a designated area. These eggs are incubated and hatched in commercial ranching facilities. From these hatchlings, some 2,000 marked juveniles, aged 8–10 months, are returned to the 'wild' on an annual basis. However, the majority of them are held back, grown to kill size and their skins (bearing the hallmark of CITES approval) sold on the international

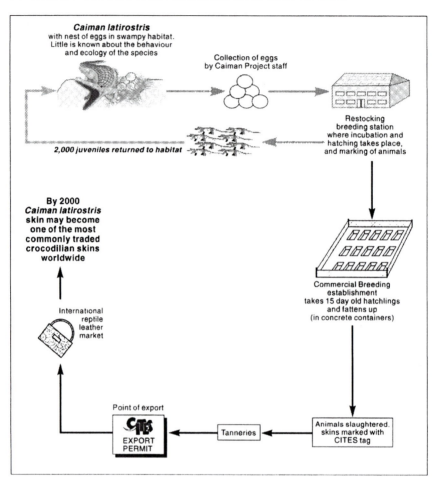

Figure 2.6 Becoming *Caiman latirostris* in the networks of sustainable use

'exotic' leather market to be made into fashion accessories (see Thorbjarnarson, 1999).

It is difficult to know, but nonetheless important to ask, what might be the impact of nest-robbing over time on the social bonds and practices, such as those of parenting, for the animals designated *Caiman latirostris*?[8] Likewise, considering the impoverished world of the hatchling, fed and raised in concrete containers with none of the creature contacts or environmental stimulii of crocodilean life, the high mortality recorded for ranched hatchlings in other contexts (Revol, 1995) should come as little suprise. For all its intentions otherwise, the animals encrypted as *Caiman latirostris* and mobilized in the websites, expert reports and conference resolutions of global conservation find themselves circulating in the flesh from Milan to

Tokyo in the performance of a wildlife network no less pernicious for the creatures embodying the wild than the Roman *venatio*.

coming home to the wild

> If wildness can stop being (just) out there and start being (also) in here, if it can start being as humane as it is natural, then perhaps we can get on with the unending task of struggling to live rightly in the world. (William Cronon, 1995: 90)

As I hinted at the beginning of the chapter, the enterprise of tracing of networks and bodies is implicitly concerned with shifting the moral geographies of wildlife from the utopian confines of the sanctuary or ark of wilderness. I return to this theme in closing, picking up where Cronon's essay 'The trouble with wilderness' leaves off – a place that is hard to imagine even though (or perhaps because) 'we' inhabit it already. How does re-cognizing the place of the wild on the 'inside' of this shared dwelling-place raise the ethical standing of the animals (and plants) which inhabit this designation? Does it render the designation wild redundant as a marker of species and spaces worthy of considerability? These are not easy questions to answer, but several preliminary points can be made which are taken forward in the next chapter.

I have sought to demonstrate some of the insights to be gained from a more symmetrical analysis of the ways in which animal lives are caught up in, and matter to, the performance of two very different orderings of wildlife. Table 2.1 summarizes some of the key lines of comparison that have been drawn between the wildlife networks of the Roman *venatio* and the species inventories of wildlife management today. Their juxtaposition has served 'to bring what seems far away close up' (Probyn, 1996: 13), opening up the chronological distance between the two networks to unsettle the space–time coordinates of the wild as the spatial and bodily remnants of a pristine past. It is a juxtaposition that also points up some uncomfortable parallels in the moral geographies of these networks which inhere in the codified and routinised practices of *venationes* and species inventories. My argument is that these geographies are not the province of some distant discourse or elevated judgement but the collective habitation of all those party, in many different guises, to *leopardus* or *Caiman latirostris* coming into being.

The geographies charted in my navigations of these wildlife networks begin by treating the living creatures that become *leopardus* and *Caiman latirostris* in and through them as if they matter. They matter as active agents who make a difference to the ways in which these heterogeneous social networks take and hold their shape. They matter as sensible creatures who are subjects as well as objects in these networks. And they matter

Table 2.1 Performing the wild

Empire	Rome (Civilization)	Science (Evolution)
Heterotopic sites	Games arenas	Animal bodies
Network vernacular	Military discipline	Expert management
Ordering technologies	Hunting and combat	Classification and conservation
Ordering aesthetic	Living spectacle	Species inventory
Currency of wildlife	Blood	Genes

analytically to the ways in which we make sense of the practical orderings of social (human and animal) life. If they are deemed not to matter in any of these senses, then, like the *bestiarii* in Rome and the human specimens in the inventories of Life Science today, belonging to the species *homo sapiens* will prove small comfort against the abuses sanctioned towards such 'non-persons'. This is a faultline that no amount of reinforcement of universalist notions of human rights will heal over and which, as we have been reminded with sanguine regularity throughout the genocidal twentieth century, will always be prised open as long as the monstrous category of the animal can be mobilized as grounds enough for treating someone as 'less than human' (Bauman, 1996; Finkielkraut, 2001). To re-place the wild topologically is to recognize that the heterogeneous social performance of wildlife configures 'human' as much as 'animal' categories and lives in intimate and precarious ways (see Sheehan and Sosna, 1991; Ham and Senior, 1997).

Like the anthropologist Tim Ingold (1988a: 15), I would argue that animals are best considered as strange persons, rather than familiar or exotic things. Their presence in the heterogeneous networks of everyday social life is multidimensional – corporeal, creative and consequential. But making their presence felt in our accounts of the social presents serious epistemological and practical problems that science (social and natural) has barely begun to admit (see Haraway, 1989; Ritvo, 1995; Glendinning, 1998). The efforts to glimpse the animal experience of becoming *leopardus* and *Caiman latirostris* rendered in this account take their impulse rather from works of the imagination, like the uncomfortable explorations of the hum/an/imal borderlands in the fictional worlds of Franz Kafka's *Metamorphosis* (1988/1905), Beat Sterchi's *The cow* (1988), Marge Piercy's *Body of glass* (1992), or the fragile life drawings of Joseph Beuys (see The Royal Academy of Arts, 1999).

The narrative strategy adopted here follows Michel Serres' (1985) injunction to attend to the fully sensate world of smell and sound, touch and taste, as a means of reconnecting human and animal experiences and diminishing the cognitive and metaphorical privilege accorded to sight as the elevated seat of human sensibility. In enjoining author and reader to engage their own animal senses in an effort to 'imagine' another's, this strategy does not pretend to circumvent the recalcitrant philosophical problems of Deleuze and Guattari's enterprise of 'becoming animal' (see Hardt, 1993) or of Serres' endeavour to conjure flesh in words (see Assad, 1999: 80). Rather, it insists with Marian Schotmeijer that these problems do nothing to diminish the claim that 'animals have sufficient Being to disturb human complacency' (1997: 140) even in the face of the engrained ways in which such claims have been rendered unutterable, let alone answerable, in the scientific calculi that pervade public life and which consistently reduce ethical questions about what counts to empirical questions about what can be counted.

It seems all the more imperative then to challenge the authority of such calculi in the governance of the non-human world by resisting the comforting urge to caricature 'Science' as the ready-made villain of the piece, even as some scientists would defend that authority by conjuring cartoon 'enemies' of their own in the familiar figure of 'anthropomorphism' or, more recently, 'Postmodernism' (see, for example, Soulé and Lease, 1995). The manoeuvres made here towards a more performative conception of wildlife, are informed by diverse theoretical efforts in the social sciences, particularly those loosely aligned experimental tacks outlined in the introduction. The theory and practice of the natural sciences are likewise variegated. There is much to learn from those working on animal sociability, semiotics and consciousness by scientists who take their creature-subjects seriously, not merely as biological specimens but as social beings (for example, Keller, 1983; Griffin, 1992; Dolins, 1999), and those engaged at the interface of bio-philosophy in an effort to retrieve the effectivity of the organism from the haystack of genes; cells and populations that have become the preferred units of biological analysis (for example, Webster and Goodwin, 1996; Rose, 1997; Lewontin, 1998).[9]

The designation 'wild' seems not to have served its animal inhabitants well, figuring them as the currency of various human desires whose value rises with distance. Even as they are caught up in the assemblage of global regulatory networks designed to 'protect' them, they find themselves objectified again in the urgent business of 'wildlife management' (Kaufman and Mallory, 1993). As this chapter has sought to show, the determination to fix the wild in the geographical and bodily spaces of animals untouched by history is intellectually and practically unsustainable. It proceeds by effacing the ceaseless intertwinings of human–animal lives that, as anthropologists (for example, Noske, 1989) and archaeologists (for example,

Clutton-Brock, 1989) persist in reminding us, haunt the places we inhabit and those 'we' do not but which are, or have been, places that others call home (Elder *et al.*, 1998). It is a spatial imaginary which has helped to deprive us of a language of connection, or kinship, beyond the 'human' and the basis for more relational ethical practices of the kind explored in the final chapter.

In practical terms however, contemporary geographies of wildlife have become too bound up with a cartographic heritage of species distribution and density, along an axis between abundance and extinction, to abandon the designation 'wild' as a strategic site in environmental politics. Protection, triggered by rarity (calibrated as endangerment), is spatially configured in the forms of areal segregation (nature reserves) or bodily confinement (captive breeding), or both. The task of revitalizing the cultural freight of wildlife in ways which enhance the well-being of animals who inhabit it is thus a thoroughly geographical one. The topological conception of wildlife proposed here marks a contribution to renegotiating the political contours of environmentalism by releasing the spaces of wildlife from the cordon of exteriority into the multiple spaces and fluid ecologies of performative networks. In this spatial imaginary the wildlife sanctuary, as nature reserve, still has a place but it is no longer one of last resort, or without a past. Rather, such sites mark one kind of dwelling-place in which to configure human–animal relations in ways which take account of the social habits and ecological orderings of all their inhabitants (Shepard, 1997).

Reconfiguring the wild on the 'inside' connects such strategic places to the myriad everyday encounters and negotiations between human and animal lives – in cities and gardens as well as forests and deserts – that sustain, rather than simply destroy, the meaning and well-being of wildlife. It reminds us that the 'wild' is not confined to the creatures and (always unpeopled) spaces of television wildlife programmes and that even here, perhaps most of all here, the 'expert' re-orderings of these always already inhabited ecologies in the networks of science, commerce and governance are a deeply political, and rightly contested, business.

3

Embodying the Wild:
tales of becoming elephant

As accustomed as we have become to the idea of a science that
'constructs', 'fashions', or 'produces' its objects, the fact still
remains that, after all the controversies, the sciences seem to have
discovered a world that came into being without men [*sic*] and
without sciences. (Bruno Latour, 1996: 23)

animal disturbances

The coinage of constructionism to which Bruno Latour refers is a familiar
currency in geographical writing appearing in many different guises from
Marxist-inspired accounts of the capitalist transition from 'first' to 'second'
nature to cultural and historical readings of the representational practices
and politics of landscape. Indeed, the relationship between the social and
the natural has been claimed as the hallmark of Geography's disciplinary
distinctiveness both as a unique meeting ground between social and natural
scientific enterprises and as the basis of human geography's particularity
within the social sciences (Massey, 1999b). As I suggested in chapter 1,
these longstanding geographical concerns have been refreshed recently by
diverse and creative traffic between human geography and science studies
communities, particularly on the frequencies of ANT and feminist analysis,
in ways which have begun to unsettle the society–nature binary and render
the tired antinomy between constructionist and realist accounts redundant
(Demeritt, 1998).

Two impulses generated by these exchanges seem to me be among the
most significant. The first of these is the re-imagination of social space
inspired by the spectre of globalization in which new spatial metaphors,
like networks, topologies and folds attempt to hold on to the situatedness
of social practices and relationships however long their reach (Hetherington, 1997c; Shields; 1997; Murdoch 1998). The second is the re-
imagination of social agency in order to recognize the creative presence of

non-humans in the fabric of social life and to register their part in our accounts of the world (Bingham, 1996; Hinchliffe, 1999). These questions of 'acting at a distance' and 'non-human agency', in the language of ANT, have become intimately related through a focus on the proliferation of socio-technologies, like computer-mediated communications or scientific instrumentation, as the silent partners in socio-material networks from international money markets to natural history film-making (Thrift and Olds, 1996; Davies, 1999). In short, they have tended to be worked through a recognition of the non-human mediators of social agency in terms primarily of the panoply of devices, inventions and inscriptions that pass through our hands, extend our presence in space–time and, for all their social effectiveness and capricious potencies, owe 'us' their coming into being (see also, Canguilhem, 1996).[1]

In this chapter I want to explore some of the tensions between this technical inflection in the ways in which non-human agency has been taken up in ANT, as a distribution of socio-material competences and effects through actant-networks (see, for example, Law, 1986; Callon, 1992; Latour 1999b) and the more visceral preoccupations of feminist analyses with the corporeal configuration of energies and elements particularized in the experiential fabric of diverse living beings (see, for example, Keller, 1983; Haraway, 1989; Hayles, 1999). The tensions between the notions and spaces of social agency mobilized in these modes of enquiry is neatly encapsulated in the dilemmas of an automated personal rapid transit system for Paris that never gets to see the light of day, as described in Latour's romance *Aramis* (1996). Contemplating its own unbecoming, Aramis asks:

> What is a self? The intersection of all the sets of acts carried out in its name'. But is that intersection full or empty? . . . How can I become a being, an object, a thing – finally a self, yes, a full set, saturated with being . . . the 'I' that humans receive at birth [when] I do not yet have a body? (Latour, 1996: 201)

Both modes of enquiry share a relational conception of social agency and acknowledge embodiment as integral to the unstable fabric of subjectivity, but their respective emphases on material configuration and experiential being frame the political and ethical import of the question 'what is a self' very differently. My intervention in these debates focuses, purposively, on non-human animals as creatures 'saturated with being' but which have been thoroughly excluded from conventional humanist notions of the subject and which sit uneasily with the extended casting of social agency figured by ANT in the guise of 'quasi-objects' (Latour, 1993) or material artefacts (Hetherington, 1997b). In this, I want to take up what Jennifer Wolch and Jody Emel, in their book *Animal geographies*, call the 'animal moment' (1998: 22), a belated recognition of the place of non-human creatures in the fabric of social life and of the legacy of their absence from social theory.[2] This ghostly populace occupies what Donna Haraway has called

> the empty spaces of both the 'culture of no culture' of self-invisible technoscientists and the 'nature of no nature' of the chimerical entitities emerging from the world-constructed-as-laboratory. (1997: 269)

Remapping and reinhabiting such spaces, as she goes on to suggest, requires 'new practices of witnessing', practices which attend more closely to the multi-sensual business of becoming animal – a relational process in which animal subjects are configured through particular social bonds, bodily comportments and life habits that are complicated, but neither originated nor erased, by the various ways in which they may be enmeshed in the categorical and practical orderings of people.

My aim in this chapter, then, is to trace the redistribution of subjectivity admitted, if rather differently persued, in ANT and feminist science studies by journeying in the company of creatures positioned at the crux of Latour's Modern paradox (1993), simultaneously the touchstones of 'a world that came into being without men . . . and sciences' *and* the objects of intensive surveillance and regulation in the name of conservation, namely wildlife. In the previous chapter I argued for an understanding of wildlife as a relational achievement spun between people and animals, plants and soils, documents and devices in heterogeneous social networks which are performed in and through multiple places and fluid ecologies – what I called topologies of wildlife. Here, I want to focus attention on the distribution of the effects and shifting positionalities of animals in and through particular spatial formations of wildlife exchange (SFWE).

The term SFWE combines the analytical impulses of John Law's notion of 'modes of ordering' as the more than narrative measure, performance and embodiment of organizational relations (1994: 20) and Nigel Thrift's notion of 'spatial formations' which emphasizes that the sociology of these ordering networks is thoroughly situated in as numerous intersecting spatial practices (1996: 47). It also seeks to enlarge the repertoire of contemporary wildlife networks and forms of exchange from the commercial nexus of activities associated with the (illegal) trade in 'exotic' flora and fauna, and the live creatures and body parts associated with that trade, to those assembled in the name of science or conservation, and trafficking in animals' biological, informatic and spectacular properties in forms as diverse as frozen sperm, genetic codes and wildlife tourism. These themes are worked through a closer scrutiny of the business of becoming elephant; a creature so long caught up in social networks of livelihood and transport, commerce and war, ceremony and entertainment that traces of its presence litter the histories and geographies of civilizations and everyday lives in several continents.

The chapter explores two contemporary SFWEs in which African elephants (alongside many other creatures) are mobilized ostensibly 'for their own good' in global networks of conservation/science as the taxanomic species *Loxodonta africana*.[3] The first network is configured through records and exchanges of computerized information on the lineages and breeding properties of animals held in zoological collections world-wide which are the calculus of efforts to 'maintain viable *ex-situ* populations' in perpetuity, particularly of animal species endangered in their native habitats. The second network is configured through an international programme of 'science-based conservation research' projects on *in-situ* species and habitats, which harness paying volunteers, corporate donors and field scientists to wildlife (and other) expeditions.

The stories I tell of these networks focus on the modes and spatialities of their practical orderings and on the ways in which these patternings work through the bodies of elephants, both in the sense of their energies and properties being variously transduced in the performance of these social networks and of their positionality as experiential subjects being reconfigured in the process. To emphasize the multiple spatialities through which elephants are mobilized the chapter traces three simultaneous moments in their presence in each network: as virtual bodies circulating in computer programmes and internet sites as digitized data or portraits; as bodies in place situating encounters in the zoo enclosure or game reserve; and as living spaces embodying the senses and relations that configure experiential subjects.

tales of foresight

This first tale concerns a wildlife network engaged in the maintenance of 'wild' animals in what are called in the parlance of conservation science *ex-situ* sites, that is in places outside the ecological complexes and social relations associated with their native habitats – so-called 'captive wildlife populations' (Earnhardt *et al.*, 1995: 493). I begin it not with the animals that have been mobilized in these networks of enclosure, but with 'a computer-based information system for wild animal species in captivity' – namely the International Species Information System (ISIS) (ISIS@worldzoo.org).[4] The tale is situated in one particular zoo that utilizes ISIS on a daily basis, and whose Director is currently serving on the ISIS Board of Trustees, namely Paignton Zoological and Botanical Garden, located in the English Riviera on the Devon coast in which Duchess, an African elephant, has been resident since 1977.

The organizational practices of these zoological agencies concern the public display and systematic record of a scientific vision of wildlife. However, a stronger social patterning in the ISIS/Paignton Zoo wildlife network is the monitoring, programming and circulation of codified knowledges about the bloodline and genetic profile of captive animals to determine optimal breeding strategies and population trajectories. This alignment of organizational practices towards the management of wildlife futures is characterized here as a mode of ordering of *foresight* which resonates with the institutional imaginary of zoo professionals. As Paignton Zoos's newly-appointed Scientific Officer expressed it:

> I think most modern zoos sort of see themselves as modern-day arks preserving the species, because for a lot of species that is going to be the only way that they will survive and hopefully at some time in the future we can reintroduce them to somewhere because something will change. If nothing changes, they are in zoos for the rest of however long . . . so to that end that's why all the stud books and the EEPs [Europaische Erhaltungs Programs] are set up because to maintain the genetically viable population you have got to avoid in-breeding and the loss of genetic information. (AP, 22/9/97)

virtual bodies

The ISIS database derives from the animal keepers' records of some 500 member zoological institutions world-wide; inputted by curators, record-keepers or registrars; and collated at the ISIS head office in Apple Valley, Minnesota, USA. The database holds biological data on age, sex, parentage, location of birth and circumstances of death of over one million individual 'specimens'. This data is made available in different formats through several software programs developed and upgraded by ISIS programmers, and licensed for use by ISIS members to inform the management and planning of zoo 'collections' in a variety of ways. The ARKS (Animal Records Keeping System) program systematizes the codification of every zoo specimen, regularizing in-house record-keeping and facilitating the exchange of information between zoos, for example, to access other zoo holdings as potential sources for animals to loan or exchange. SPARKS (Single Population Analysis and Records Keeping System) codifies data useful for the development of stud books on the captive population of particular (usually endangered) species world-wide, and the software facilitates demographic modelling and genetic management of a species' population.

At Paignton a single and somewhat temperamental computer terminal in the office of the Curator of mammals, which is tucked away behind the public cafeteria, is the zoo's ISIS interface. From here, in moments snatched in a busy working week, standardized data are entered on the life history of each specimen with individual animals being coded in several ways to reduce the possibility of errors in the system. The Curator is primarily responsible for record- and stud book-keeping at the zoo. He demonstrated his use of the system by calling up records which he had previously inputted on a mammal in which he has a particular personal interest.

> . . . You can do a specimen report which is the individual so you key up the specimen like that and you ask for its number, say like 1008 which is a golden lion tamarind. . . . Enter its sex; house name, which is our in-house name – we based this on Desmond Morris's classification of mammals in that F is primate – 100 is Tamarind, Golden Lion Tamarind, and 8 is the eighth individual which we have had here, either arrived or born at the zoo; international stud book number; the IDs of the sires and the dames – they are the ISIS numbers as recorded here; and whether it was parent reared. Now tamarinds are actually all on loan to us from the Brazilian Government . . . and he died, this one actually hit a window which was quite unfortunate. (NB, 12/12/97)

Thus mobilized as vital statistics, in terms of their parentage, fecundity, genetic profile and so forth, the manipulation and correlation of these virtual bodies on computer screens and printouts might translate into an 'optimal pairing' of creatures living continents apart. Such digitized knowledges facilitate an unprecedented managerial capacity in the practices of captive breeding. ARKS, for example, '. . . actually tells me how many females in the different age classes based on a gestation period of 180 days. It then tells you the fecundity of those females on a scale of 0–1' (NB, 12/12/97).

Animals circulating in this fragmentary way are traceable to living creatures only by means of numeric codes tattooed or tagged on their bodies, or pet names that carry from keepers to software. These virtual coordinates give and hold the shape of a creature through the network and are particularly significant in the mapping and modelling of captive populations of endangered species. For example, SPARKS ranks animals according to the desirability of their breeding attributes:

> Basically what the programme does at the end of the day is that it lists all the males and females in the living population in its, what is called mean kinships, so the degree of relatedness of one animal to all the others in the population, so if one is say a wild caught animal which has never bred before it is going to be very high up the list, if it's produced several offspring then it has a lot more animals related to it, so it will come down the list as opposed and it will come down further than say a non-wild caught animal that's only got two relatives, that has yet to breed itself will come high up on the list. So that's where the trained eye will come in and see whether the computer is lying or there is some jinx crept in to it. But it ranks the thing and then you are supposed to, ideally, I mean it doesn't always work out for group animals, but certainly in the case of pairs then you want to try and pair on the level, to put high ranking with high ranking, and that's what we do in the orang-utans in two weeks time. (NB, 12/12/97)

Report date : 02 Mar 1997   Loxodonta africana - african elephant   Page 7

Current world population

STUD ID	S	SIRE ID	DAM ID	IDENTIFICATION Type in brackets	DATE OF ARRIVAL dd/mm/yy	LOCATION Terms in brackets		HOUSE NAME	ARKS ID	FCOEP	SUB-SPECIES
LOCATION: OPELZOO-KRONBER (0. 1. 0)											
7902	F	WILD	WILD	( )	17/09/81?	OPELZOO KRONBER	(P)	aruba		0.000	
LOCATION: PAIGNTON (0. 1. 0)											
7005	F	WILD	WILD	( )	04/04/77?	PAIGNTON	(P)	duchess	200	0.999	
LOCATION: PEAUGRES (2. 2. 0)											
6905	F	WILD	WILD	( )	27/07/73?	RAMAT GAN	(?)	ettie	730011	0.000	knocherhauri
					13/08/93	PEAUGRES	(P)	ettie			
8703	M	7402	7702	( )	19/02/87	RAMAT GAN	(B)	junior	870011	0.125	knochenhaure
					02/08/92	PEAUGRES	(P)	junior			
8801	M	JOHA	JOHA	( )	12/06/90	PEAUGRES	(?)	johntie		0.000	
8902	F	7402	6401	( )	30/04/89	RAMAT GAN	(B)	yosepha/bahaton	890015	0.250	knochenhauer
					02/08/92	PEAUGRES	(P)	yosepha			

Figure 3.1 Duchess's entry in the European inventory of *Loxodonta Africana* (European EEP for *Loxodonta africana*, 1997)

The European EEP for *Loxodonta africana*, for example, covers some 150 elephants and 38 institutions and is maintained by a species coordinator at the Zoological Center Tel Aviv-Ramat Gan. Here, Paignton Zoo's Duchess is present in her ARKS identity, code number 200 (see figure 3.1).

But in the labour-intensive business of maintaining species' stud books, the cartographies of wildlife are complicated by various, and sometimes competing, regional calculi based primarily in Europe, North America and Australasia and a convoluted pattern of 'line management' over breeding decisions distributed through several international conservation agencies. As the Scientific Officer at Paignton Zoo explains:

> . . . the European system runs two levels of management at the moment, there's the really intensively managed species which are the ones that are the most critically endangered in the wild, and they are managed by full stud books and the European stud book coordinator has actually absolute power and can say to any zoo 'do that with this animal', some of the zoos ignore them but, and the aim of that is to conserve 95% of the genetic availability for 200 years. So all the population genetics is calculated to do that. And then there's the lower level and that's the, the high level is EEP and there is actually only a European endangered species programme for it, that's what that is, and then there is the

European stud book and the stud book coordinators of that don't have
absolute power, they can only make suggestions and that is basically run
to avoid in-breeding, its not quite such a strict genetic control . . .
maximum avoidance of in-breeding I think is the terminology. (AP,
22/9/97)

At a major meeting in Easter 1997, member zoos discussed the next
phase of ISIS/REGASP – a regional collection planning capacity facilitating
the international coordination of genetically viable populations, the software for which has been available since March 1996. But the rivalry of
regional scientific traditions and logistics of coordination aside, it is
apparent that this plethora of virtual transactions – the currency of the
zoo–ark ideal – is rarely able to translate its prized 'genetically viable'
offspring beyond the enclosures of captive-breeding.

<u>bodies in place</u>

For all its calculated foresight, then, this is a wildlife network whose
precious cargo is destined, for the most part, to remain firmly in the hold.
The virtual lines of force of the ISIS network which configure the zoo as a
conservation site are always brought to earth by the physical fabric of the
zoo as a showcase for public entertainment and education, which is
designed to keep animals and people in their proper place. Paignton Zoo
prides itself on being the only combined zoological and botanical garden in
the UK at the present time. Its publicity stresses the zoo's conservation and
scientific role and, under its current Director, Paignton has been in the
vanguard of translating this repositioning into the physical organization of
the zoo space in the form of an Environmental Park. The design of animal
enclosures is informed by published behavioural research and keepers'
more intimate knowledges of animals' habits, which feed into curatorial
concerns with the everyday management of bodies in place (not least with
an eye to breeding). But the layout and arrangement of animal enclosures in
the space of the zoo are at least as forcefully shaped by the passing
spectatorial sensibilities of those who keep the turnstiles in motion as by
those of their permanent inhabitants.

Under the reorganization of its 75 acres as an 'environmental park',
Paignton Zoo is currently developing six 'habitats', starting with a Devon
Woodland habitat and expanding to Forest, Wetland, Desert, Savannah
and Rainforest habitats. Over time, many of the existing specimens and
groups in the zoo's animal collection will be moved out of their segregated
taxonomic enclosures and geographic clusters and rehoused in their appropriate habitat. The aim is to give the visitor the impression of seeing an
animal in its 'natural' context, including the presence of other appropriate
species. As the zoo's Director explains, 'We have got to try and create an

illusion, that's what we are trying to do, that they walk into these areas and they have the illusion of being in a savannah or forest or whatever' (PS, 24/9/97).

This £6 million re-modelling of the zoo's material space and the repositioning of animal exhibits within in it (see figure 3.2) is being funded by a European Regional Development Fund with a public education agenda. But it also resonates with a levelling-out of the wildlife landscape projected in this SFWE and described by Paignton Zoo's Director as a closer integration of 'captive' and 'wild' spaces in the global network of wildlife science and conservation with '. . . zoos taking animals out of cages and putting them into bigger areas, national parks becoming ever increasingly smaller and managing their animals much more intensively . . . generally it's the same principles and they are merging, coming closer together' (PS, 24/9/97). Enrolling the public into this patterning of foresight is a fraught but vital task because visitor entrance fees remain the zoo's principle source of revenue to meet the costs of animal upkeep, capital maintenance, staff employment and conservation commitments.

> Well, nothing's free, there are no free lunches in this game, that's for certain. And with regard to the ISIS, I don't know what its costing us, sort of £1500, £2000 a year or something like that, its all based upon the size of the collection and as well as how big you are, its done on a scale.

Figure 3.2 Paignton Zoo Environmental Park brochure (1997)

In addition to that, that's not the only cost, the direct cost of actually making use of the software, people make a contribution to running that software . . . you have got to attend meetings, you make a contribution as toward the running of it, someone has got to do it, and only zoos are going to do it. So we are directly contributing to . . . the management groups that manage the populations, the Taxon Advisory Groups and we also are members of the international Conservation Breeding Specialist Group, a sub-group of IUCN [International Union for the Conservation of Nature]. (PS, 24/9/97)

One of the spaces at Paignton Zoo that is most actively mobilized in this public enrolment is the elephant enclosure inhabited by Duchess and her sole companion for the last 20 years, an Indian elephant called Gay. Feeding time at the elephant enclosure is one of the main visitor attractions in which the keepers and elephants put on a daily performance to an amplified commentary, an event which has been harnessed recently to raising funds for a field research project on the Nigerian Forest Elephant in which the zoo is involved. At the same time, this interweaving of 'ex-situ' and 'in-situ' bodies and spaces in the display practices of the zoo lays the ground for refashioning the material space of Duchess and Gay and their translocation to the new 'savannah' enclosure. As a re-placing of Paignton's elephants, the savannah enclosure also materializes the patterning of foresight. With 87 per cent of *Loxodonta africana* holdings recorded in the European EEP coming from the wild, the coordinator has recommended that

> if breeding is a goal of the EEP, then changes must be made in management. Some of these changes are now occurring regarding concepts of handling, monitoring and in the construction of the new facilities that are being built. There is still much to be done to complete the picture, and improve the lives of African elephants in captivity. (Terkel, 1996)

It is to the living space of Paignton's African elephant, Duchess, that attention turns next.

living space

The mode of ordering of foresight can be traced to the most intimate spaces of wild-life – its multi-sensual living space. Duchess has been the only *Loxodonta africana* in Paignton Zoo's collection since 1977, the female calf of 'wild' parentage (sire and dame). This much can be gleaned from her vital statistics recorded in the European EEP listing for 1996 (see figure 3.1 above). According to the Curator of mammals, she was acquired from the much larger elephant collection at Longleat, one of Britain's first 'wildlife parks'. Duchess and Gay have shared the same concrete enclosure since their arrival at Paignton. Her only other social interactions in this space have been with keepers and, at a distance, with the passing public and spectators at feeding time. Whereas most animals in the collection have a fairly remote relationship with their keepers,

> . . . with something like elephants, because it is one of the few animals that have human contact and need it as well, there is only a certain few of us that actually do this. . . . Because it takes a while for the elephants to get to know someone new and it takes a while for you to get to know them and you have to sort of build up this sort of bond between you and the elephant because if you don't, you have got to work safely in there and if you can trust them because they are so strong and so powerful if you can't trust them . . . you know. . . . If they don't accept you then that's it, you are off, you can't do it. (J, elephant keeper, 23/9/97)

Her trunk, mouth and feet are checked first thing each day by the keepers for signs of ill-health, but the 'highlight' of her daily routine centres on feeding time, which is a public event timetabled for two o'clock in the afternoon (see figure 3.3). The keeper's commentary tells the assembled crowd that Duchess consumes 60 bananas, 40–50 apples, a crate of cabbage and a bucket of special feed mix every day. She responds to signals from her keeper to make choreographed gestures – lifting a foot, opening her mouth or displaying her trunk finger – to the delight of the audience. But, like the keepers, as a spectator one is always aware of Duchess's size and the uncertainty of her compliance in this balance of powers in the public eye.[5]

This routinized living space changed dramatically in September 1997 with the reorganization of the zoo. The elephants were the first species to be moved to the new 'savannah' enclosure. The move was a moment (or, as it turned out, 24 hours) of high anxiety for the zoo staff, particularly the elephant keepers and the Curator of mammals, because neither Duchess nor Gay had ever been out of their old enclosure. These anxieties were heightened by the presence of a BBC television production unit filming the whole procedure for broadcast in a series about zookeepers.[6] The elephant keeper was most concerned about the animals experiencing panic:

J: I mean what really concerns us is the fact they will start to panic and they can sort of hurt themselves. Well, this is it, you know we can generally calm them down, I mean, we have had a few panic attacks off them when we have sort of been practising this hobbling. . . .

SW: And what are the signs of panic? How do you read it?

J: Roarings, trying to snap this chain that we have got them on, kicking out, I mean just going into a panic. . . . We have managed to . . . get them back under control, sort of soothe them.

SW: How do you do that?

J: Just talking to them, sort of.

Figure 3.3 Feeding time at the elephant enclosure, Paignton Zoo

SW: They recognise your voice?

J: Hand in their mouth, stroking their tongue, just to let them know that its all right, we are here, you know. (23/9/97)

The new enclosure is in part designed to improve the elephants' everyday lives; loose sand to make feeding a more 'interesting' foraging experience; a wallow and shower, even though the keeper was doubtful they would be used as Duchess and Gay always hide indoors when it rains. But their keeper seems keenly aware that if this new environment is to 'stimulate' them in the manner foreseen, it would only do so with practice – by the elephants inhabiting that patterning of foresight.

> All the things like the dead trees and all that in there, that we have got in the new place. . . . that they can climb up in the mornings . . . so the elephants can sort of feed natually rather than from the ground, make them stretch up to feed. Whether it will work or not we don't know, they just might lean up and pull the branch down and feed . . . the way they usually do. We don't know, but at least we can sort of give them the option. (JM, 23/9/97)

On arrival at the new savannah habitat both animals remained within the elephant house, refusing to go outside and 'explore' their new enclosure. Ordinarily this act would restrict the elephants from the public gaze, but the new elephant house is architect-designed for internal viewing, with a raised public gallery so that they can see more of what goes on behind the scenes and view the elephants even when they are sheltering from the weather.

Judging by the experience of the September move, the stress and disorientation that Duchess might experience by any potential re-introduction to the wild would be traumatic. Taxonomically, she certainly belongs to *Loxodonta africana*, but the elephant she has become through her life at Paignton Zoo bears only distant relation to those of her kind at home in the African bush, even as such living spaces are themselves being increasingly reconfigured in the same patterning of foresight in which she is caught up.

tales of authenticity

My second tale concerns a wildlife network which harnesses the energies of popular interests in wildlife conservation to the performance of scientific research 'in-situ', that is within the ecological complexes and social relations associated with animals' native habitats. The public face of this SFWE, at least until January 1998, was Earthwatch, an organization responsible for recruiting paying volunteers to these projects – the so-called EarthCorps.[7] Its organizational 'twin', the Centre for Field Research (CFR), was established shortly after Earthwatch's inception in a Boston pub 25 years ago. The CFR is the vetting and grant-awarding agency for research proposals from the scientific community to run projects. A programme of some 140 Earthwatch field projects in 50 countries world-wide is approved each year in consultations between the CFR and Earthwatch organizations in the USA, the UK, Australia and Japan. Here, the focus is on Earthwatch Europe and the CFR, which work from the same office in Oxford, UK. Earthwatch Europe attracts funding from a variety of sources: the Earthwatch membership and 'expedition' fees of volunteers (55 per cent), and sponsorship from corporate (30 per cent) and institutional (15 per cent) donors. Again, we journey through this wildlife network with the African elephant, this time those mobilized in an Earthwatch field project on the 'Okovango elephants' in northern Botswana during the 1990s.

The organizational practices of these charitable conservation agencies promote participation in scientific projects as a demonstration of 'global citizenship' under the slogan 'now you can help save the planet' (www.earthwatch.org). However, the most potent social patterning in the Earthwatch/CFR wildlife network is the expedition experience – making it 'real' both in the sense of servicing proper science and of encountering wild(er)ness first hand. This alignment of organizational practices towards the configuration of wildlife as experience is characterized here as a mode of ordering of *authenticity*, echoing Earthwatch's own framing of expedition as 'an experience you will never forget' (*Earthwatch*, 1997) and 'a journey that may change your life' (welcome letter to EarthCorps volunteers, undated).

virtual bodies

Earthwatch has been described in the travel pages of a British broadsheet newspaper as an 'adventure holiday' organization (*The Guardian*, 'An awfully big adventure', 20/5/97), but the organization itself eschews this ecotourism label: 'We are not a travel organisation, we are a charity and somehow we need to find a way to differentiate ourselves from superficially similar commercial organisations' (EW, Marketing Director, 12/11/97). An organizational renaming as the 'Earthwatch Institute', which came into effect in its 1998 literature, has been one way to address this problem.[8] At the same time it is a self-aware appeal to the organizational culture of the scientific community. Recruiting high-calibre scientists to its programme has always been important for Earthwatch.

> What really gives us the prestige and authority are first of all our science advisers and trustees, and when they see who our Chairman is and they see who our science advisers are that instantly sort of will calm any fears they might have that we might be just another organization or whatever. And then also the quality of the people who are leading the field projects. Often not necessarily the individuals but the institutions the individuals come from. So, if for instance, we were to tell people that our project in Cameroon is done in collaboration with Kew Gardens that will obviously give it some kind of authority. (RB, Development Director, 11/11/97)

In a context where institutional credentials are so key, the organizational change of name asserts scientific authenticity; prestige, authority and seriousness of purpose are all freighted by the 'Institute'. This is, after all, a global network for the advancement of conservation *science*.

Recruitment to this network is through the colourful printed and electronic mediators of the annual Earthwatch brochure and website, in which the virtual bodies of diverse creatures, plants and places are mobilized in the programme of Earthwatch field projects across the globe – from bees and orchids in Brazil to the forests of Bohemia. The point of entry for this elephant tale is the 1997 glossy, large-format catalogue in which a range of marketing devices are deployed to attract volunteers, characterized as 'ordinary' members of the 'general public' who can pay something in the region of US$1,000–2,000 to join one of the advertized expeditions for a week or a fortnight (plus their own travel costs to and from the field site). These marketing practices are finely tuned – the visual and textual vernacular is carefully honed to a public that is closely monitored for its social

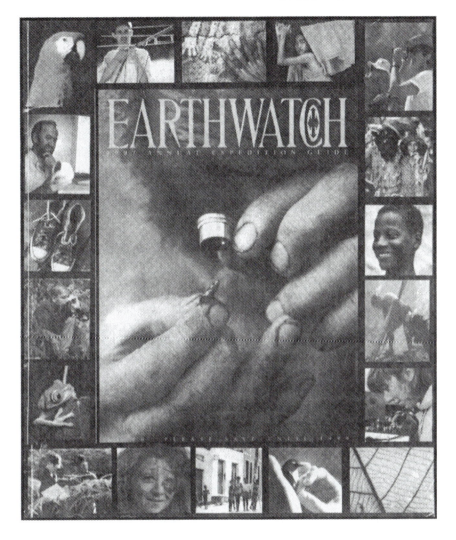

Figure 3.4 Earthwatch *Annual expedition guide* (1997)

profile, habits and values via Earthwatch membership procedures. In the case of wildlife projects, animals are mobilized as photographic portraits of individuals or groups, close-up or in scenery evocative of the promise of an authentic experience of 'the wild' (see figure 3.4). As the Marketing Director puts it:

> We tried then to deal with each project in turn by firstly presenting it in a very attractive well-designed format with a large photo which you know again could be slightly sort of travel orientated in terms of either focusing on the lovely place or in the case of animals obviously sort of the cuteness factor, but we have tried to offset that by always showing sort of shots of

people working and everything else to make it, to break down the fact that you know this isn't just a lovely photo but you can actually be in that photo. (EW, 12/11/97)

Like other creatures, elephants circulate as virtual bodies in the guise of these electronic and print portraits standing for others of their kind and for the radical otherness of wildlife and the places it inhabits. These virtual bodies are the most potent figures in the enrolment of the paying volunteers that make the net-work. But their powers are unequally distributed with some portraits, like those of elephants, acting as what are known in the trade as, 'flagship species' or 'charismatic mega-fauna'. From a marketing point of view,

> . . . its again one of the projects where my work is simple. We publish the catalogue, we include the project and we include a picture of an elephant, people will sign up for it. Its one of the ones where it would hardly be worth my time or money to even advertise it because its such a flagship species and people are so interested in it that you know, my work is done for me, people will sign up straight from the catalogue. (EW, 12/11/97)

At the same time these virtual bodies insinuate themselves into the scientific calculus of Earthwatch field projects, complicating the clean-cut parameters of the 'rigorous peer-review process' which this SFWE is at pains to be seen to practise. In a very real sense, it is these animal portraits that translate between the 'hard science' credentials of the project leaders and the 'inexpert' enthusiasms of the paying volunteers. In this act of translation, animal portraits connect very different wildlife passions and curiosities by conjuring authenticity as a common currency of 'being there' and 'getting your hands dirty' in the fleeting space–time of the field site. Here expertise is diffused in the mundane business of practical research:

> . . . what we have tried to say in the front matter and the body of the text is that really you don't need any skills to join an Earthwatch project, that everything you do need to know will be taught to you by the scientist once you are actually out in the field and also the majority of skills are basic data collection and it's surprisingly easy, in fact almost too easy, to the point of being very boring. (EW, 12/11/97)

Moreover, the programme of projects circulated in the brochure and website arises from a coordination of marketing and scientific priorities and practices. All projects considered for entry in the annual expeditions listing will have had to pass the CFR's review protocols. Potentially, these involve any of the 11,000 names in the computer database of scientific referees held at Oxford in the grading of project proposals although, in practice, most traffic passes through the extensive personal network of the Scientific Officer. The final selection of projects from the CFR's approved list

balances their scientific merits against marketing and financial considerations. Weighting the volunteer costings of projects mobilizing flagship
species is one of the mechanisms available to Earthwatch to cross-subsidize
less 'charismatic' projects, and one of the factors influencing the final
profile of the expedition brochure. Here, the animal portraits carrying the
promise of an authentic wildlife experience across time and space become
calculable in dollar digits.

<u>bodies in place</u>

If animal portraits are the virtual wildlife currency between scientists and
volunteers in this SFWE, their coinage holds only in so far as the promised
immediacy of wildlife experience is realized in the field site – in other
words, through encounters with material bodies in place. The advertised
project must deliver these for the reputation of the Earthwatch/CFR
network to endure and its regiment of volunteers to be reinforced year on
year (some 35–40 per cent of volunteers sign up for a second project). In
1990, a biologist from Oxford University submitted a project proposal on
'Elephants and their habitats in Northern Botswana' to the CFR, identifying its scientific coordinates as the intersection of Biology, Conservation
and Ecology. The field sites for the proposed expedition were located in the
Moremi wildlife reserve and Chobe National Park. The objective of the
research was to gain empirical data about elephants' diets and their
interaction with vegetation which could fuel a model to predict the long-
term impacts of elephants on woodland habitats to guide conservation
management.

The project focused on various intermediaries to trace elephants in
place, counting dung balls in the field sites to calculate the number of
elephants per unit area and markers like broken branches to detect their
consumption of woody vegetation at a range of sampling sites (see figure
3.5). Over time, monitoring of plant regrowth and seedling establishment
would indicate the resilience of vegetation. In this way, the study redistributed the agency of the elephant through the mapping and counting of traces
of its metabolic presence.

The project's scientific credentials rested on obtaining large numbers of
samples over a relatively long period of time. Its routine, relatively
unskilled research tasks perfectly suited the project to the Earthwatch
profile. Teams of volunteers could perform the basic activities associated
with transect sampling, gathering biomass and leaf litter data. The proposed timing of the project, from July to September, also matched the 'ice-
cream business' seasonality of Earthwatch Europe's programme and
volunteering patterns. The project was initially approved by the CFR for a

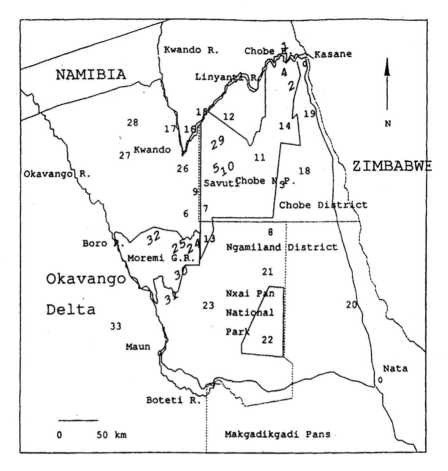

Figure 3.5 Earthcorps briefing map of 'Okavango Elephant' project field sites (Earthcorps briefing pack for Okavango Elephant project, 1993)

three-year period (August 1991–93), favourable to the kind of longitudinal analysis needed in a region susceptible to extended periods of drought.

The 'Okavango Elephant' project, as it was repackaged for the Earthwatch brochure, did not in fact focus directly on the material bodies of elephants. But the principal investigator gave some assurances of the 'contact' so critical to the authenticity of this wildlife network, in a welcoming letter to potential volunteers for the second phase of the research (1994–96) which drew on experience from the first phase.

> While in the field, the tediousness of work and rough living conditions were often compensated by the occasional sighting of elephants, lions and buffalo (much to the PI's demise!). Many more sightings of the diverse wildlife in the reserves were waiting at the end of a long African

day during game drives. At night-time, around the camp fire we often observed the reflections of light from the eyes of hyenas that prowled around the camp site. (undated)

As it turned out, the 1994 season had to be abandoned because the principal investigator was charged, and seriously injured, by an elephant in the field. The following year was, for various reasons, to be the last for the 'Okavango Elephant' project. In the final report, the principal investigator drew attention to the following research findings.

The decline in elephant densities in most plots in the period between the dry seasons of 1992 and 1993 was coupled with a general decline in the frequency of recent elephant damage to plants (–, 1994). However, the increase in elephant densities in plots from 1993 to 1995 was not clearly evident in terms of increased damage to plants. Furthermore, there was no support for a previously derived correlation which described an increase in elephant densities and accompanied decline in the abundance of tall trees (–, 1994). (undated) (– author's name removed)

In providing valid scientific data on bodies in place, volunteers energies in collecting dung and foliage may not have been quite the 'first-hand' wildlife experience they had in mind. But the elephant encounter which played a part in bringing the project to a close endures keenly in the collective memory of this wildlife network.

living space

A small group of people (6–8 volunteers per team, three teams per year) brought together for a short space of time (12 days) to assemble elephant dung and foliage in the parched landscape of northern Botswana in the dry season embodies the EarthCorps experience. Conditions described (from the point of view of the scientists and volunteers) as remote and harsh expose this 'global citizenry' to human diseases like sleeping sickness and malaria, which are endemic to the area, and to the heat, flies and dust that permeate the rhythms of the southern African day for all its inhabitants. The mode of ordering of authenticity patterns not only the marketing and scientific activities of this SFWE but, to this extent at least, calibrates the living spaces of elephants and humans in a common animal currency of bodily vulnerabilities – the urgencies of water and food, protection from the sun and wind, and rest.

For the elephants themselves, however, whose eating habits are pieced together with such intent, the patterning of authenticity is more tellingly insinuated into their very mortality. As the principal investigator's welcoming letter to potential volunteers on the 'Okavango Elephant' project put it:

There is concern . . . that woodland habitats could carry only a limited number of elephants. An excess number of elephants could induce an irreversible damage to the woodlands through the overutilization of vegetation. Little information is known, however, about the capacity of the region to sustain the elephant population. Such information is vital to the local Department of Wildlife and National Parks for the development of future management policies. (undated)

A subsequent project, 'Botswana's elephants', listed in the 1997 Earthwatch brochure under the leadership of a different scientist, puts the issue (and lives) at stake even more starkly.

Without a serious poaching problem and without culling, Botswana's elephant herd has swelled to roughly 75,000, the largest in Africa. Chobe National Park . . . harbours an estimated 30,000 elephants in its 11,000 square kilometers. Can the park support that many elephants? Park officials are leery of culling, because they fear the surviving elephants will become shy of tourists, a chief source of foreign currency. But during the dry season . . . when the grasses wither, elephants turn to browse, especially to acacia trees. Elephants strip them of bark, break branches, and push them over. The damage is especially severe around waterholes, where large numbers of elephants congregate. . . . Your work here is vital to resolving two major controversies in sub-Saharan Africa – about the wisdom and efficacy of digging more waterholes (to attract more wildlife and, hence, more tourists) and of culling elephants. (Earthwatch, 1997: 39)

Here, the agency of elephants is admitted into this SFWE at its most potent: as active architects of the landscape; as thoroughly social herd creatures whose energies are distributed through their relations with others – tourists, animals, plants and the fabric of soil and water – that configure their living space. Ironically, this potency is carried through the Earthwatch wildlife network by the close-up image of a charging elephant making a trail of dust – an image shot by the leader of the ill-fated 'Okavango Elephant' project (see figure 3.6). The brochure caption reads 'You won't get this close to elephants you'll watch at waterholes in Botswana's Chobe National Park. But you'll be close enough to feel like you're this close' (Earthwatch, 1997: 39).

Once again, the balance of powers circulating in this patterning of authenticity entangles a host of agencies – of elephants and acacias, volunteers and park managers, dung transects and computer models – that refuse the easy distinctions between the 'natural' and the 'cultural' even in this 'wilderness'. But the nomadic habits, social bonds and metabolic needs of the creatures whose living spaces are the subject of such close scrutiny are

Figure 3.6 Botswana's elephants: an Earthwatch 'wildlife experience' (Earthwatch *Annual expedition guide*, 1997)

increasingly being translated here, and in other social networks of conservation and tourism, into abstract units of consumption, drained of all the
multi-sensual business of becoming elephant. Even as their potencies nourish
the authenticity of this wildlife experience, they are being calculated as
excessive, a problem of living space that needs to be cut down to size.

wildlife in performance

> Instead of a single space–time, we will generate as many spaces and times as there are types of relations. Thus, progressing along jungle trails will not produce the same space–times as moving along [transport] networks. . . . The difference between these trips . . . comes from the number of others one has to take into account, and from the nature of these others. Are they well-aligned intermediaries making no fuss and no history, thus allowing smooth passage, or full mediators defining paths and fates of their own? (Latour, 1997a: 174–5)

These two tales, of foresight and authenticity, have much to say in response to the question of 'what is a self'. They complicate the hyphenated sociologies of actant-networks and spatialites of socio-technical orderings by introducing the disruptive figure of the animal. Travelling in the company of the African elephant through these networks of wildlife exchange animals can be seen to be simultaneously mobilized (set in motion) as 'well-aligned intermediaries', the virtual bodies circulating as the abstract units of wildlife data and icons 'making no fuss and no history', *and* emotive (moving on their own account) as 'full mediators', the sensible and experiential inhabitants of 'paths and fates of their own'. Moreover, the relationship between these very different mobilizations is configured through yet another fold in the social fabric of these wildlife networks, the situating of animal bodies in place.

The two SFWEs explored here illustrate the diffusion of network enrolments beyond the tidy designations of 'ex-situ' and 'in-situ' which align their organizational practices with a bio-geography that pervades the imaginary and infrastructure of global wildlife conservation (Jeffries, 1997). This bio-geography territorializes distinctions between the natural and cultural, the wild and the captive, which are increasingly undermined in the practices of conservation/science as zoos redesign their exhibition spaces with (at least half) an eye towards captive breeding, and protected wildlife in reserves and parks is managed ever more intensively. But just as surely, this folding, or pleating, of network spaces also redistributes animal subjectivity in terms of their experiential repertoire and social bonds in ways which complicate the business of becoming elephant. Zoo animals like 'Duchess' and the 30,000 or so in the herds of Chobe National Park may be kindred under the taxon *Loxodonta africana*, but in many other senses they are worlds apart. For all the scrutiny, vetinary intervention and population management, the elephants of Chobe still lead nomadic, socially rich and ecologically complex lives. For all the attention to design, stimulation and care in her new savannah enclosure at Paignton Zoo, Duchess has become habituated to a more impoverished repertoire of sociability, movement and life skills that will always set her apart.

The modes and spaces of wildlife performance in these two networks seem to me to illuminate the tensions between the ANT configuration of agency as the distribution of material competences and the insistence of feminist science studies on the situatedness of radically different kinds of subjects within these networks, including that of the story-tellers (Elam, 1999). This exploration of the distribution of social agency in the SFWEs of ISIS/Paignton zoo and Earthwatch/Okavango, in a way which endeavours to acknowledge the positionalities of the elephants caught up in them, can certainly be read as affirmation of Latour's sense of subjectivity as 'a circulating capacity, something that is partially gained or lost by hooking up to certain bodies of practices' (1999b: 23). But this does not seem enough. The important question, as Annemarie Mol suggests in relation to her work on the distribution of social agency in medical practice, is whether the crucial moments in this distribution 'are not those where 'patients' act as an agent, but rather those where they (we) are defined, measured, observed, listened to, or otherwise *enacted*? (Mol, 1999: 87).[9] This is a question, I would contend, that cannot be answered without greater attention to the diversity and particularity of subjugations of many very differently embodied kinds that complicates the 'they (we)', as well as the 'act/*enacted*', moments of this complex ontological politics. This attention to the material-discursive practices in which being is configured or, as Karen Barad has put it, in which agential realism is sedimented, focuses attention on the 'real consequences, interventions, creative possibilities, and responsibilities of interacting within the world' (1999: 8). The multiple ways in which elephants are on the move in the spatial formations of wildlife exchange explored here open up these facets of the too often flat political and ethical landscapes of ANT.

Section 2

. - . . . -

> It is a vital concern of every State not only to vanquish nomadism but
> to control migrations and, more generally, to establish a zone of
> rights over an entire 'exterior' over all the flows traversing the
> ecumenon. (Deleuze and Guattari, 1988: 385)

Those of us several generations removed from the land have grown so
accustomed to its proprietorial re-mapping that the vernacular of
'property' now routinely confuses land itself with the specific legal
designs of exclusive forms of ownership (Hann, 1998). For the majority
of the world's population whose livelihoods are still bound up with its
energetic vicissitudes, the distinction remains as obvious as it is vital. But
some sense of the intimate violence of private property rights as a mode
of (dis)possession is pricking even western complacencies, as we witness
new forms of bodily fragmentation and reattribution in the clamour for
intellectual property rights (IPR) in cells and genes, not least those of
'our' own species, attending the hyperbolic inventiveness of the post-
genomic life sciences.[1] As efforts to render life amenable to appropriation
as 'biological resources' threaten to mesmerize us with their apparent
novelty, it is timely to recall that they observe the syntax of enclosure
that has characterized modern political ordering since the 'mythic charter
of the demon-king of modernity' (Fitzpatrick, 1992: 73) – Thomas
Hobbes' *Leviathan* (1985 (1651)). At its starkest, the bio-political
evocation of this parable of the commons today entertains two rival
impulses: the reconstitution of the subjects and spaces of political
community through the architecture of human rights and global
commons; and the de/re-territorializations of corporeal difference within
and between human and other kinds through the technical and legal
assemblage of new forms of bodily commodity (see Petchesky, 1995).

Property is one of, if not, *the* primary currency of ongoing
conversations between Law and Geography (see Blomley *et al.*, 2001).
This should come as no suprise given their shared complicity in the
cartographies of governance, commerce and science and the recalibration
of a litany of 'exterior' and/or 'prior' space–times within the coordinates
of modernity's compelling embrace, whether those associated with the age

of empire or, today, with the rubric of global environmental management. Critical explorations of this 'law–space nexus', as Nick Blomley describes it (1994), have variously insisted that property has less to do with defining 'the relationship between a person and his [sic] things, than with the relationship that arises between persons with respect to things' (Ackerman, 1977: 26).[2] Such work usefully highlights the socio-spatial orderings effected through property rights and the spatial imaginaries that pervade legal criteria and argumentation in determining evidence of entitlement. In particular, it has variously focused attention on the role of property law in (re-)fashioning social divisions and identities in terms of 'classes' through the axiom of labour (e.g. Tribe, 1978); 'races' through the metric of civilization (e.g. Pagden, 1987); and 'genders' through the register of marriage (e.g. Ferguson, 1992). But I want to suggest that, however inadvertently, the emphatic focus on the 'persons' in such accounts has occasioned a tendency to neglect the significance of the 'things' transacting relations between them and, more significantly, to evacuate what David Delaney calls 'the physicality of the law' (2001). In other words, while such accounts take neither the letter of the law nor the jurisdictions it inscribes as self-evident configurations of justice or space, they do little to unsettle the distinctions it reiterates between the social and the material, the human and the non-human.

In the essays in this section, I want to extend the critical conversations between geography and law by interrogating the 'zone of rights' governing de/re-territorializations of the body politic in terms of the ways in which law maps the flesh as well as the earth, complicating the bio-geographies of belonging and exchange through which persons and things, interiorities and exteriorities, are con-figured. Rather than treating the slippage between the legal devices of deeds or patents and the land or artefacts to which they pertain as a simple category mistake, I take up Marilyn Strathern's injunction to attend more carefully to the legal significance of 'the capacity to body forth the effects of creativity' (1999a: 134) in the fabric of 'things' – that is to substantiate proprietorial assemblages. Such 'capacities', I suggest, are distributed achievements, articulating the exertions and affordances of heterogeneous bodies and elements whose hybrid performance is partially and provisionally fixed through the devices of property law (see also Battaglia, 1994). For example, the spatial codification of 'real' property as a grid-like surface finitely divisible into mutually exclusive estates is both unimaginable and impracticable if we substitute the socio-materialities of land for those of air or water (see Wiel, 1934). Likewise, the practices of patent law are being metamorphosed by the socio-materialities of 'living things', as their promiscuous corporealities exceed the modality of inventiveness configured by/as wrought and machinic artefacts (Correa, 1995).

My argument in these chapters is that legal practices and their durable incarnation in property devices like deeds and patents are just as significant as those of science and technology in 'making the cut' between the social and the material, through their determination of who/what constitutes persons and things (*persona* and *res*). As 'the basis of modern law' (Fitzpatrick, 1992: 82), property can be seen to play a crucial part in governing the shifting coordinates of the 'inside' and 'outside' of political community, territorialized in the nation state and configuring who counts as a political (and proprietorial) subject, and of 'society' and 'nature' territorialized as distinct ontological domains, and configuring what constitutes an object of property right. Not only are science and law, technology and property, invariably complicit in effecting such distinctions (see Black, 1998), but they exhibit compelling parallels in their modes of ordering. For example, both are characterized by a universalizing ambition that fashions the world as a *terra incognita* or *terra nullius*, to which they alone bring order by effacing or subsuming all other modes of knowledge or regulation. Furthermore, law, like science, is inclined to efface its own practices, masquerading its fabrications as self-evident accomplishments. It is vulnerable to critical scrutiny only by getting up close and tracing its (un)making through the laborious assemblage of intrepretative communities, ritual words and phrases, documentary precedents and professional protocols; performative achievements that are always partial, contestable and incomplete.

Enjoining Deleuze and Guattari's sense of the spatial and temporal turbulence of governing socio-material flows as/through a 'zone of rights', these chapters explore two provisional moments in the geo-political contours of proprietorial enclosure. In navigating the knotty assemblage of these events, I follow the documents that are both artefacts and mediators of the talkative fabric of legal conduct. Chapter 4 interrogates the European 'settlement' of Australia in the wake of the so-called Mabo ruling by the Australian High Court in 1992, which overturned the legal doctrine of '*terra nullius*' on which the continent's territorial governance as both colony and Commonwealth had been premised. It traces continuities and tensions in proprietorial justifications for the exclusion of Aboriginal peoples from these bodies politic through parliamentary debates in Britain in the 1830s, contemporaneous with the passage of the South Australia Colonization Bill, and in Australia in the 1990s during the passage of native title legislation. Chapter 5 explores the governance of flows and spaces between nation states through the institutions of global environmental management, focusing on the assemblage of plant genetic resources (PGR) as a 'heritage of humankind'. It charts the contested efforts of the United Nation's Food and Agriculture Organization to establish an International Undertaking on Plant Genetic

Resources to regulate the de/re-territorialization of beneficial rights in PGR against the monopoly impulse of western forms of intellectual property rights and corporate appropriation.

4

Unsettling Australia:
wormholes in territorial governance

Those people . . . who having fertile countries, disdain to cultivate the earth and choose rather to live by rapine, are wanting to themselves, and deserve to be exterminated as savage and pernicious beasts. (Emmerich de Vattel, 1916/1760: 37)

It is obviously more difficult to shoot noble savages than people who were no better than animals, who roved over the landscape like so much nuisance fauna. (Noel Pearson, Director Cape York Land Council, 1993a)

first encounters

On 22 August 1770, 12,000 miles from home, Captain James Cook planted a Union Jack on a small island in the Torres Strait off the northern tip of a land mass then known in Europe as New Holland. This gesture, on the aptly named Possession Island, staked the British Crown's first claim to sovereignty over the island's continental neighbour against the colonizing impulses of rival nations (Scott, 1940). On 3 June 1992 four plaintiffs from the adjacent island of Mer, annexed by Queensland in 1879 on the authority of Letters Patent issued by Queen Victoria, won their ten-year battle in the Australian High Court to overturn this claim and to assert 'that the Meriam people are entitled as against the whole world to possession, occupation, use and enjoyment of the land of the Murray Islands' (ALR, 1992: 499).

Such fragile messengers of territorial governance, a flag and a document, have had volcanic repercussions for the constitution of Australia. At the crux of their significance lies the arcane legal doctrine of *terra nullius*, literally no one's land (Simpson, 1993). Inscribed in the still inchoate body of eighteenth-century international law, *terra nullius* legitimized the annexation of 'uninhabited lands' by settlement as an acknowledged

means, alongside conquest and secession, for the proper conduct of colonization by 'civilized' nations. Articulated here are two allied but different
moments of possession: *dominium*, which vests absolute jurisdiction over a
territory in the political authority of sovereign nation states, and *ius*, which
accords property rights of various kinds to the sovereign subjects of those
states. In the self-effacing facticity of legal terminology, 'the law' appears as
the simple arbiter of competing territorial claims between preconstituted
states and subjects. In practice, as the (un)settling of Australia makes clear,
the assemblage of the Law as an autonomous and unified domain is both
an outcome and an instrument of these peculiarly modern forms of political
authority, agency and territory coming into being (Fitzpatrick, 1992).

The indigenous peoples of Australia, like so many elsewhere, whose
social orderings differed from those emerging in Europe, were rendered
commensurable with colonial ambitions by being cast out of the social
altogether as 'primitives', 'savages' or 'barbarians' (see Pagden, 1987;
Greenblatt, 1991). Only in this way can the British assumption of sovereignty with the arrival of the first fleet in New South Wales, and the
subsequent vestiture of proprietary titles under English common law, be
squared with the evidence in the letters, diaries and official reports of those
who arrived in Cook's wake, that the land they encountered was manifestly
not uninhabited.[1] As the dialogue staged at the start of this chapter
between de Vattel, one of the most influential authorities in international
law in the eighteenth century, and Noel Pearson, a leader of the Aboriginal
land rights movement in Australia today, suggests, *terra nullius*, for all its
Latinate decorum, has been the viaduct for a barbarization of indigenous
peoples consonant with, rather than merely against the grain of, the law
(Rose, 1984). A cartoon in the *Aboriginal Law Bulletin* shortly after the
Mabo judgment makes this deceit at the heart of the settler-nation brutally
graphic (see figure 4.1).

The spaces of colonization were evacuated, literally and metaphor-
ically, by the practices of European settlement which incorporated the
inconvenient people they encountered into the fabric of the land they
coveted as inhabitants of the 'state of nature'. This secular myth was
promulgated in political, legal and, later, scientific discourses which recali-
brated spatial differences in the socio-material organization of life as
temporal stages in a universal progression of 'mankind' such that the
spaces encountered beyond Europe became comprehensible as spaces
before Europe (McClintock, 1994). Early modern political theories of the
rise of 'civic' as against 'natural man', like Thomas Hobbes' primal
covenant of governance (Tuck, 1979) and John Locke's commonwealth of
proprietors (Tully, 1993), furnished new rationales for European colonization in the seventeenth and eighteenth centuries by shifting this 'state of
nature' from a Godly estate of divine moral order to an offensive waste in

Figure 4.1 *Terra nullius* I: the violence of 'settlement' (cartoonist: Andrew Ireland; *Sydney Morning Herald*; reproduced from *Aboriginal Law Bulletin*, 1993, 3/62: 10)

the jurisdiction of enlightenment reason (Hulme, 1990). Scientific theories in the nineteenth century, most notably those associated with evolution, honed that reason into a still more potent instrument of colonization, removing the 'state of nature' to an outdoor museum of remnant species destined for extinction (Macgregor, 1993). If North America had been the archetype of Locke's 'state of nature' (Arnell, 1996), Australia was to epitomize its evolutionary variant (Kuper, 1988) and the fictional, but all too consequential, transposition of those who lived there from 'noble savages' to 'nuisance fauna'.

The diverse indigenous peoples of Australia found themselves transfixed by their prior presence in the shifting cosmologies of European colonization at the junction between history and biology as a relic and thoroughly distant human kind (Reece, 1987; Béteille, 1998). This rendering of Aboriginality as a categorical incarnation of 'stone age peoples' or 'the lowest of the savage races' was enthusiastically fleshed out in various scientific enterprises, most notably that of anthropology, until well into the twentieth century (Anderson K., 1999, 2000). But its reach went much beyond their conventional curiosities, colouring the bureaucracies of colonial and commonwealth administration and the facts and fictions of European self-regard.[2] Thus, for example, the Director of the Anthropological Institute in London at the height of empire espoused the view that 'we may suppose that they [Australian Aborigines] represent one of the earliest stages in the progress of mankind towards that high culture which is exhibited by the European' (Stanisland Wake, 1872: 84). Some 60 years later they appeared in a popular fortnightly serial *The Science of Life*, this time as one of a number of 'divergent or retrograde forms of human life . . . the sociological equivalents of the platypus and echidna . . . not ancestral survivals but side branches' (H.G.Wells *et al.*, 1931: 864). I do not rehearse such racist calculi to evoke an historical moment in the European discernment of Australia or to rake over the practices of colonial science, but rather to bring home their insidious familiarity in the geographies of the 'post-colonial' present – like the itch that harbingers a cold-sore (see Willems-Braun, 1997; Livingstone, 1998). My purpose is to begin to work against the spatial and temporal parameters of such accounts of Australia which, by virtue of their own logic, would have us consign them to the safe distance of long ago and far away.

The 1992 Australian High Court judgment, or Mabo ruling as it has become known, after Eddie Mabo, one of the Torres Strait Islanders who brought the case, made just such an effort to 'put the past behind us' in its opening statement.

> The facts as known today do not fit the 'absence of law' or 'barbarian' theory underpinning the colonial reception of the common law of England in its relation to indigenous people. As the basis of the theory is false in fact and now unacceptable in our society the Court should not allow the common law to be, or to be seen to be, frozen in an age of racial discrimination. (ALR, 1992: 409)

Even as the tidy assertions of Crown sovereignty over the Australian colonies belied a welter of contested, and often bloody, practices of land appropriation, occupation and use in the late eighteenth and nineteenth centuries (see Reynolds, 1982, 1992), so the neat about turn in the Mabo ruling at the close of the twentieth settled nothing. Rather, as Australian commentators have been the first to recognize, this landmark judgment opens up new and no less disputed configurations of the 'commonwealth' which Australia aspires to be.[3] In this chapter I want to map some of the continuities and circumlocutions in the territorial governance of Australia which disconcert the coordinates of Colonies and Commonwealth, settlers and aborigines, waste and cultivation and make space for lags, pleats and lacunae in its geographies of dis/possession. As well as countenancing the commotion of passages – of families and transports, migrants and minerals, tourists and soap operas – that thread through and against such binary cartographies of the Australian body politic, these 'wormholes' admit my own journeys into this strangely familiar continent as visible traces in this account.

If it was the ghost of Captain Maconochie that first conjured Australia in my geographical imagination, it was the historian Paul Carter's book *The road to Botany Bay* (1987) which focused my attention. It was one of those books that became a rite of passage for graduate students of my generation and one which made Australia travel in ways few academic texts have done before or since. This self-styled 'spatial history' traces the practices of place-naming – explorations, surveys, maps, inventories – which he argues brought Australia into being, as against simply ascertaining something which was already there. For Carter, this kind of history 'begins and ends in language' (1987: xxiii) and the portable inscriptions which aligned the land to their makers' passing intent. His account of Geography as a praxis of earth-writing found a warm reception in a discipline just then beginning an energetic engagement with the politics of representation, and which later solidified into something of an orthodoxy in cultural geography (see Thrift, 1991). The kinds of geographies that move me, then and now, do not sit comfortably in such prohibitively meaningful worlds but they prefigured my first journey to Australia some five years later in important ways. Not least, it was at an early conference exploring these approaches at University College London in 1987 that I first met Jane Jacobs, newly arrived from Australia to start her doctoral research here, whose work and friendship drew me closer to her country.[4]

Like Carter, I want to map the unsettling of Australia not as a process of un-making history 'where the past has been settled even more effectively than the country' (1987: xx), but of history continually in the making. In the Bergsonian terms which have so influenced such alternative historicities, this means acknowledging 'the continuous progress of the past which gnaws into the future and which swells as it advances' (Bergson, 1983/1907: 498). But I want also to counter their shared tendency to treat space as an immobilizing dimension in which the infinitely creative business of history is brought to earth and somehow rendered inert. If territoriality is to be appreciated in performative terms, then this repudiation of history as a uni-directional sequence of events has to be matched with a repudiation of geography as a uni-versal plane on which such a history unfurls (see

Massey, 1999a). The hectic patterns of very different kinds of comings and goings, encounters and separations, occupations and dwellings in which territoriality inheres, fold space as well as time in multiple, relational and provisional configurations. In these complicated space–times there are, as Nigel Thrift has put it, 'no stable and complete orders, only tentative and fractional orderings' (1999: 302).

Where Carter's spatial history provides clear directions to Botany Bay, there are few signposts to Wagga Wagga or dual places, like Uluru/Ayer's Rock, and no bearings at all for the places that non-European (even non-British) migrants have made their home. In other words, it maps an Australia in which the colonial spatiality of discovery/naming becomes omnipresent, categorically displacing 'Aborigines' to a nostalgic hunter-gatherer past and confining 'migrants' to the terrain of British settlement (see also Gelder and Jacobs, 1995). Part of the problem I think is precisely Carter's insistence on the primacy of language and his pre-occupation with naming as a definitive spatial practice, which mimic the colonial impulse to empty the world out and render it legible in European terms.[5] I shift emphasis here to other spatial practices in the constitution of Australia, notably those of sovereignty and property, in which the word is a no less potent but much more uncertain and situated agent of territorial governance. Most obviously, the polished formalities of the letter of the law and parliamentary statute permit as much as they restrict the latitude for (re)inscribing territorializations. But, just as significantly, percolating through territorialities of any kind are more visceral, fraught and tacit practices which connect bodies and soils, identities and places, in ways that disrupt the scripted geometries of dis/possession and admit more heterogeneous geographies of belonging.

In this chapter I interrogate conflicting efforts to renegotiate the colonial settlement of Australia as it configures 'the land' and 'the native' in the fabric of the nation. I work this interrogation through an interweaving of three documentary moments in the territorialization of Australia: the seven judgments (one dissenting) comprising the 1992 Australian High Court (Mabo) ruling itself; the Australian parliamentary debates (in the House of Representatives and the Senate) which culminated in the Native Title Act of 1993; and the 1836–37 British House of Commons Select Committee Report on Aborigines (British settlements) which deliberated the impact of settlement on Aboriginal peoples in the midst of Australian colonization.[6] I am less interested in the textuality of these documentary moments than in their fallibility as mediators in the assemblage of sovereignty and property and the cartographies of the nation which they freight. My purpose is to highlight the eddies and fissures in the territorialization of socio-material relations which overspill and undermine the borders of their scriptual intent.

unquiet lands: (dis)locations in the body of the law

> [We need to get] rid of the assumption that the ownership of land naturally breaks itself up into estates, conceived as creatures of inherent legal principle. (Justices Dean and Gauldron, ALR, 1992: 441, para. G)

> In discharging its duty to declare the common law of Australia, this court is not free to adopt rules that accord with contemporary notions of justice and human rights if their adoption would fracture the skeleton of principle which gives the body of our law its shape and internal consistency. (Justice Brennan, ALR, 1992: 416, para. E)

My first visit to Australia (October–November 1993) coincided with the frenzied political climax of the Government's legislative response to the Mabo judgment in which the ruling Labor Party, in uneasy alliance with a handful of Democrats and Greens and against a bitter Opposition coalition of Liberal and National Parties, steered its Native Title Bill on to the statute book just before Christmas in 1993 (see figure 4.2). This legislative process, like the Mabo judgment itself, was mindful of international as well as national agendas. Prime Minister Paul Keating used the Australian launch of the UN International Year for the World's Indigenous People in Redfern, Sydney in December 1992 to peg Australia's laggardly record on indigenous land rights and governance to those of other settler-nations (Aboriginal Law Bulletin, 1993).

Based in Melbourne and travelling by car and plane around Victoria, New South Wales and ACT (Australia Capital Territory), sometimes in the company of Jane Jacobs, I became keenly aware of a land that exceeds its settlement. One journey in particular brought this home forcefully. Returning to Canberra to catch a plane back to Melbourne after a weekend spent in the spectacular Blue Mountains north-west of Sydney, replete with heritage sites and tourist trails, we decided on impulse to take the 'scenic route' to the airport. Within an hour the orderly tarmac came to an abrupt end and the comfort of our hirecar succumbed to the dust and craters, bends and forks of a dirt track. A two-hour motorway trip turned into a six-hour cross-country epic that bore no relation to the distance on the map and reduced us to nervous silence as our flight time came and went and an unlit darkness fell.

During these travels, talk of Mabo and native title was everywhere but Aboriginal voices were confined in my hearing to the television, radio and newspapers. This reinforced an immediate and enduring impression from the documentary records of legal and parliamentary proceedings that the rhetorical evocation of Aboriginal people, who are constantly spoken about or 'for' through personal or expert European witness, resounds with their absence (see de Certeau, 1985). This ghostly effect is all too apparent

Figure 4.2 Mabo: history makes the headlines (photographer: Mike Bowers; *The Age*, 22/12/92)

in the materials I interrogate here and comes into focus in the second part of the chapter which deals with the shifting place of 'the native' in the governance of the nation. In this first section I want to deal with the other dimension of territoriality, the land itself, which by virtue of its own liveliness and the vicissitudes of possessory practices is never finally or completely settled.

Grand theories and conventional wisdoms of various kinds have cast private property in heroic terms, as a geometric mantle levelling the rude earth before the irresistable and comprehensive advance of colonialism (and/or capitalism, civilization, modernity).[7] Thus incorporated into the law, land appears synonymous with property; an isotropic plane parcelled up into discrete and mutually exclusive estates (Blomley, 1994). But as the fragments from the Mabo judgment which open this section suggest, in practice the 'lay of the land' and the 'body of the law' are neither as quiet nor as seamlessly interwoven as such accounts would have it. Rather, land is a much more energetic configuration of earth and air, water and minerals, animals and plants, as well as people than a surface area contained by lines on a map (Ingold, 1986: 147). I want to explore two currents in the more fluid possessory landscape that this documentary moment admits and which, through the political networks of parliamentary debate and legislation, were channelled in contested and provisional ways in the Native Title Act 1993. First, the performative nature of territoriality is nowhere more evident than in the convolutions of common law which infuse the Mabo judgment. Here I pick up on one of the most significant dis-locations between its transposition from England with the first fleet and its assumption of a distinctively Australian habit in the Mabo case, namely, the disturbance of the relationship between sovereignty and property as coincident parameters of 'settlement'. Secondly, the proprietory landscape is complicated by a variety of legal rights which fall short of freehold, the gold standard of private property,[8] and which configure different kinds of entitlement to various substantive qualities and capacities of the land, often in relation to the same area or plot (Simpson, 1986). I consider the implications of the majority judgment's cautious admission of this 'proprietary pluralism' for the accommodation of native title within the same universe of partial and provisional land rights as a host of 'settler' titles in Australia, from pastoral leases and mining licenses to squatter grants and game permits.

Common law as it came to be understood by the end of the eighteenth century rested on the pronouncements of judges whose proceedings subjected heterogeneous practices to rulings made 'permanent and certain' in the legal record (Fitzpatrick, 1992: 60–2).[9] But for all these efforts to regularize it as written or positive law, common law then and now bears the hallmark of 'customary' modes of justice and, as a consequence, is less conventional and more amenable to the variability of circumstance than public or statute law (Murphy, 1994). One of the most well-worn common law maxims, that 'possession is nine tenths of the law' (see Epstein, 1979), pricked the consciences of many eighteenth- and nineteenth-century English commentators witnessing the dispossession of the labouring poor by the parliamentary enclosures of the commons at home and of native peoples by the colonial 'settlement' of foreign lands.[10] Sir William Blackstone, a leading contemporary legal figure, for example, was moved to ask:

> So long as it [colonization by settlement] was confined to the stocking and cultivation of desert uninhabited countries, it kept strictly within the limits of the law of nature. But how far the seising on countries already peopled, and driving out or massacring the innocent and defenceless natives, merely because they differed from their invaders in language, in religion, in customs, in government, or in colour; how far such a conduct was consonant to nature, to reason, or to christianity, deserved well to be considered by those who have rendered their names immortal by thus civilizing mankind. (Blackstone, 1803/1765, ch. 1: 7)[11]

In the colonial context his question was largely rhetorical, although it was vigorously taken up by the House of Commons Select Committee on Aborigines (British Settlements) which held such a system of colonization to be 'based upon a principle of unrighteousness' (BPP, 1836: 515). But it becomes pivotal to the Mabo judges' declaration of the common law of Australia and is cited by the leading judge, Justice Brennan, in his ruling (ALR, 1992: 418, para. D).

The majority judges held that Aboriginal land rights had not been extinguished by the assumption of Crown sovereignty but, more than this, that these antecedent rights represented a 'burden' on the Crown from the first under the provisions of English common law. Justice Brennan (with the majority) sought repeatedly to underline that

> As the Governments of the Australian Colonies and, latterly, the Governments of the Commonwealth, States and Territories have alienated or appropriated to their own purposes most of the land in this country during the last 200 years, the Australian Aboriginal peoples have been substantially dispossessed of their traditional land. They were dispossessed by the Crown's exercise of its sovereign powers to grant land to whom it chose and to appropriate to itself the beneficial ownership of parcels of land for the Crown's purposes. . . . Their dispossession underwrote the development of the nation. . . . It is appropriate to identify [these] events . . . in order to dispel the misconception that it is the common law rather than the action of Governments which made many of the indigenous people of this country trespassers on their own land. (ALR, 1992: 434 *passim*)

For the first time in Anglo-Australian jurisprudence the Mabo judgment disentangled sovereignty and property and opened up the space between the Crown's radical title to the territory of the colony, a matter of public law, and its absolute beneficial title to the land, the jurisdiction of the common law (Edgeworth, 1994). This space also served to demonstrate the distance the High Court of Australia had come from the feudal vestiges of English law which it had inherited and to assert its independence in declaring the common law of Australia. Accounting for the majority judges' departure from previous High Court rulings on the issue, Justice Brennan went so far as to note that

> it is not immaterial to the resolution of the present problem that, since the *Australia Act* 1986 (Cth) came into operation, the law of this country is entirely free of Imperial control. The law which governs Australia is Australian law. (ALR, 1992: 416 para. F)

It is this interval between sovereignty and property that furnished the majority judges with the legal grounds to admit the existence of Aboriginal land rights into Australian common law, in circumstances where these had not been expressly extinguished by valid Crown grants or statutory acts. Vacant Crown land and, by extension, Crown land with invalid or expired third party leases or licenses, thus became subject to native title claims (see figure 4.3).[12]

Moreover, in the course of their deliberations four of the majority judges variously queried the privileged status of cultivation in European political and legal theory as the only hallmark of proprietary interest capable of recognition at common law.[13] Citing precedents from North American Courts and the International Court of Justice, Justice Toohey argued that

> It is presence amounting to occupancy which is the foundation of the title and which attracts protection, and it is that which must be proved to establish title . . . not the occupation of a particular society or way of life. (ALR, 1992: 486: para. B)

This broader definition of occupancy required proof of the meaningful use of land 'from the point of view of the members of the society [of users/claimants]' (ALR, 1992: 486 para. C). It opened the way to acknowledging more varied and robust forms of native title in accordance with traditional uses, rights and obligations, irrespective of whether or not they conformed to 'English or European modes or legal notions' (Justices Deane and Gauldron, ALR, 1992: 441 para. C).[14]

In their different ways, then, the majority judges in the Mabo case make tentative trails across the 'unbridgeable gulf' between Aboriginal and European territorial practices, belaboured by anthropologists (see Ingold, 1986) and popularized by travel writers like Bruce Chatwin (1987). Consonant with their efforts to establish an autonomous jurisprudence,

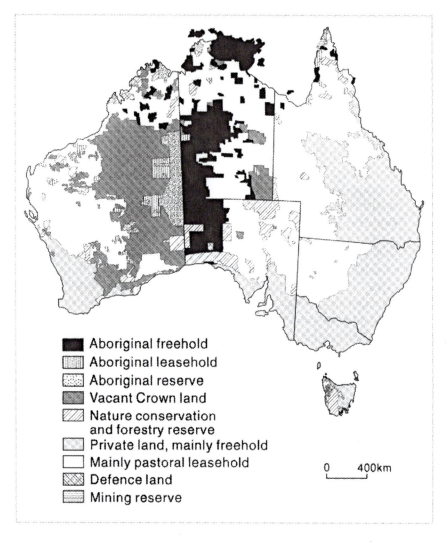

Legend:
- Aboriginal freehold
- Aboriginal leasehold
- Aboriginal reserve
- Vacant Crown land
- Nature conservation and forestry reserve
- Private land, mainly freehold
- Mainly pastoral leasehold
- Defence land
- Mining reserve

0   400km

Figure 4.3 Official map of land tenure in Post-Mabo Australia (adapted from Gelder and Jacobs, 1998: 140)

they moved cautiously towards incorporating the traditional customs of indigenous peoples into the corpus of Australian common law. But, as we saw at the start of this section, the interval between sovereignty and property also marks the junction at which the High Court judges set about plastering over the cracks that their own deliberations had exposed in the 'body of the law' by affirming Australia's territorial integrity as a sovereign nation and ruling challenges to the acquisition of Crown sovereignty outside their jurisdiction (Patton, 1996; Reynolds, 1998).

Unsurprisingly, the Mabo ruling provoked consternation among those in the mining and pastoral industries who felt their territorial interests were compromised, as well as the legislatures of States dominated by these industries and in which the majority of native title claims could be anticipated.[15] Habitual champions of 'law and order', they now impugned the High Court for indulging in 'judicial activism' and overturning 'two hundred and five years of settled land law' (see, for example, Michael Cobb (NSW, National Party), Hansard (HoR), 24/11/93: 3587; Morgan, 1992). At the same time Mabo gave unprecedented grounds for hope in the long struggle of indigenous Australians to recover their stake in the country. While this stake was strategically focused in the statutory networks of Aboriginal representation, like the Aboriginal and Torres Strait Islanders Commission (ATSIC) on advancing native land rights, it was taken by others to embrace the larger questions of self-governance and sovereignty (see, for example, Mansell, 1992; Pearson, 1993b).

The Government of the day shared the High Court's inclination to distance Australia from its colonial ties. Prime Minister Paul Keating, introducing the Native Title Bill on its first reading in Parliament, put it with characteristic verve.

> Mr Speaker, some seem to see the High Court as having just handed Australia a problem. The fact is that the High Court has handed this nation an opportunity . . . [for us to] make the Mabo decision an historic turning point: the basis of a new relationship between indigenous and other Australians. (Hansard (HoR), 16/11/93: 2877)

While seeking to 'protect native title to the maximum extent practicable' (ibid.), the Government's legislative process, unlike the proceedings of the High Court, had to negotiate the political hurdles of securing a majority in parliamentary divisions, countering the opposition of major interest groups and State legislatures, and holding on to the Labor Party's electoral support.[16] The outcome, inevitably, was a compromise which saw the 'practicability' of protection for native title restricted by stringent criteria for determining who qualified as 'native' (which are taken up below) and by assurances to other land owners that their titles were secure, which went beyond those signalled in the Mabo judgment. Those forms of property exempt from native title claims were defined in the Bill, and by the Prime Minister, at the outset.

> Only validated freehold grants, residential, commercial and pastoral or agricultural leases, and validated Crown actions basically involving public works, will extinguish Native Title. Naturally, existing reservations for the benefit of Aboriginal and Torres Strait Island people will be preserved. (Hansard (HoR), 16/11/93: 2879)

Notwithstanding these compromises, the parliamentary debates are marked by a persistent tension between Government efforts to protect legislatively the extant native title rights now recognized at common law, and a concerted Opposition strategy which sought to cast the legislation as an attempt to 'create' a new form of land right specific to Aboriginal people that threatened the proprietory interests of other Australians and the integrity of the nation.[17] The Senate Labor Party leader Gareth Evans insisted, for example, that

> I would not wish it to be thought that the existence and recognition of native title is itself a special measure, because that native title is part of the common law of the land. It is no different in its essential character from other kinds of proprietary right recognized as a matter of common law. (Hansard (Senate), 16/12/93: 5030)

Several of his colleagues in the House of Representatives, like backbencher Neville Newell (NSW, Labor Party), argued that common law native title did not threaten other land titles but could, as it already did under the statutory provisions operating in several national parks, like Uluru and Kakadu, co-exist with mining or pastoral interests.

> The native title bill is not, as the opposition would have us believe, a 'land grab' or a 'free-for-all'. It is simply a recognition under common law of the traditional rights and interests of the Aboriginal and Torres Strait Islander community. It is not a vehicle that will open Australia up to farcical land claims, nor is it a bill that will end Australia's mining or agricultural operations. There are many examples of mining operations on traditional aboriginal land where the local aboriginal people and the mining company enjoy a very harmonious relationship. (Hansard (HoR), 25/11/93: 3723)

But the supple understanding of law, land and history which such arguments, like the Mabo judgment itself, implied was met with an instinctive, and sometimes cynical, incomprehension by Opposition speakers whose interventions recite a variety of cherished certainties, three of which echo through the Hansard record like choral refrains. First, that such a 'radical new' departure in the law of the land should be the prerogative of Parliament and not the High Court. Thus, for example, Senator Short (Victoria, Liberal Party), argued that

> Overturning *terra nullius* and establishing a new form of land right, native title, regardless of whether one thinks it's a good or bad idea, is surely the proper responsibility of parliament and the people and not of the judiciary. (Hansard (Senate), 15/12/93: 4709)

Secondly, that native title legislation threatened investment in, and exploitation of, the land – the 'bedrock' of Australia's economic prosperity. Senator Calvert (Tasmania, Liberal Party), for example, charged that

Something like 330 million hectares, or about 43 percent of our total land, are held under pastoral lease. . . . The bill will provide native titleholders with the right to veto and impede the economic development of such land. This will place directly at risk some of our large export earning industries. (ibid., 15/12/93: 4695)

And thirdly, that the clock could not now be put back to rectify past injustices (however regrettable, etc.). As Senator MacGibbon (Queensland, Liberal Party), for example, observed

With the best will in the world we cannot go back and correct the wrongs of years ago. . . . Let us not bog ourselves down, flagellating ourselves about the wrongs that occurred in the past. They are beyond change; they just cannot be altered. (ibid., 15/12/93: 4704)

After all the talk of an 'historic moment' with which speaker after speaker, whether for or against the legislation, prefaced their contribution to the long hours and late nights of parliamentary debate, the final provisions of the Native Title Act assented to on 24 December 1993 pertained to less than a quarter of the land area and 7 per cent of indigenous people, whose total numbers amount to under 2 per cent of the Australian population. For its most reactionary opponents, Native Title remained a contradiction in terms. For many indigenous Australians displaced over several generations from their cultural ties with the land, it proved legally and socially remote. But, perhaps the most troubling legacy of the 1993 Act was that it reiterated the settler–native parameters of the nation even as the Government sought to shift the political terrain of post-Mabo Australia towards a more heterogeneous and fluid accommodation of indigenous and migrant passages and to realign its place in the world from a European antipodes to an Asian neighbour.

unruly subjects: (re)placing the 'native' in the nation

Most average suburban whites in this country can expect to live their lives without ever meeting one of this country's original inhabitants. They will never talk to an Aboriginal Australian, let alone know one. (Chris Pyne (South Australia, Liberal Party), Hansard (HoR), 24/11/93: 3516)

Policy makers must accept that indigenous people are not a special category of disadvantaged souls who require attention or even caring and gentleness. We are peoples with rights and imperatives of our own. (Mick Dodson, ATSIC Social Justice Commissioner, *Sydney Morning Herald*, 9/5/95)

During my second visit to Australia (April–May 1997) I was invited to Rockhampton, a rural town in tropical central Queensland, to give a seminar in the Social Sciences Faculty at the University campus there at the same time as the town was hosting 'Beef '97'. This bi-annual gathering is simultaneously a national celebration of cattle cultures and communities and an international forum for the promotion of the Australian beef industry. An 'international visitor's' pass had been arranged for me to join a group of south-east Asian government delegates at a lunch, followed by a display of 'traditional dancing' in a nearby Aboriginal cultural centre. It was a surreal and uncomfortable occasion for an out-of-place 'Pommie' woman. In the company of several, mainly Korean and Filipino, officials (and their wives) intent on breeding stock and bull semen I met for the first time a handful of the people who inhabit the designations 'aborigine' and 'pastoralist' that populate the parliamentary record. Beef '97 had also brought the campaign entourage of Pauline Hanson's 'One Nation' Party to town, peddling her white Australia invective on home turf (http://www.onenation.com.au). She was met by a vocal crowd of protest-ors, indigenous Australians among them, but there was no doubting the resonance of her message with these cattle farmers' collective sense of themselves as a beleaguered minority.[18] This place was a long way from the cosmopolitan streets of Melbourne and Sydney and the cool corridors of government in Canberra. I was left wondering whether the folk attending Beef '97 were that much more familiar to 'most average suburban whites' than the 'Aborigines' habitually cast as strangers in the nation and whose presence is measured in terms of welfare rather than rights.

Twin descendents of the colonial antonym of 'settler' and 'native', the designations 'pastoralist' and 'aborigine' continue to configure the territo-rial practices and imaginaries of Australia through their antagonistic enfolding in the land. With the magnetism of a compass their categorical polarity gives the country its bearings, reiterating the contours of belonging and the spaces of dis/possession even as the people these categories purport to describe slip their moorings or re-cognize themselves as social minor-ities.[19] How can such a culturally diverse and overwhelmingly urbanized society still be in thrall to this terrene binary? How is it articulated in the impossible polity of the 'average Australian' and with what consequences for ongoing efforts to renegotiate the place of the 'native' in the governance of the nation? I want to address these questions here by tracing (dis)continuities in the configuration of 'settler/pastoralists' and 'native/aborigines' between the deliberations of the Australian Parliament during the passage of the Native Title Act 1993 and those of the British House of Commons Select Committee on Aborigines of 1836–37, some 160 years before. This is a wormhole travelled by several Commonwealth parlia-mentarians and Mabo judges in the tracks of one of Australia's most influential historians, Henry Reynolds.

Having admitted native title into Australian common law and deter-mined that its content 'must be ascertained as a matter of fact by reference to [the traditional] laws and customs' of native titleholders predating the advent of British Settlement (ALR, 1992: 429), the High Court left the detailed business of establishing and regulating native title to the 'evolution' of case law (see Bartlett, 1993). One of the primary purposes of the 1993 Native Title Act was to put more flesh on these rudimentary common law bones. The Act defined native title as

> . . . the communal, group or individual rights and interests of Aboriginal peoples or Torres Strait Islanders in relation to land or waters where:
> (a) the rights and interests are possessed under the traditional laws acknowledged, and the traditional customs observed, by the Aboriginal peoples or Torres Strait Islanders; and
> (b) the Aboriginal peoples or Torres Strait Islanders, by those laws and customs, have a connection with the land or waters; and
> (c) the rights and interests are recognized by the common law of Australia. (Native Title Act 1993, Part 15 'definitions', clause 223 (1): 104)

At the same time, 'Aboriginal peoples' are defined in the Act as 'peoples of the Aboriginal race of Australia' (ibid., clause 253: 122) who, as numerous speakers recall, were specifically excluded from the Constitution of the Australian nation drawn up in 1901 and only gained the formal status and rights of citizenship following a public referendum in 1967.[20]

The common ground between the political parties, such as it was, lay in their acknowledgement of the widespread incidence of poverty and social exclusion among Aboriginal communities today, in terms of any number of indices of health, education and standard of living. Hansard also records occasional references to the regular audience of Aboriginal observers in the public galleries, and the influence of a close working relationship with the ATSIC on the language adopted by Government proponents of the Bill, particularly that of ministers (Dodson, 1994). But these feint Aboriginal presences in the parliamentary performance of the nation are overshadowed in the Hansard record by a more insistent spectre – racism. The 'R word', as one parliamentarian coyly refers to it, smoulders through the typeface in angry exchanges and mutual loathings and lurks behind a patina of self-censorship on the part of leading Opposition speakers who complain that their views on 'Aboriginality' have been stifled by a prevailing climate of 'political correctness' (see, for example, Tim Fischer (Leader of the Australian National Party), Hansard (HoR), 23/11/93: 3425).[21]

The Native Title Act 1993, like the Mabo ruling, explicitly rejected the 'legal social darwinism' that had consigned Aboriginal people to a position 'too low on the scale of social organisation' to constitute an effective polity. The machinery of native title tribunals set out in the Act sought to extend the evidentiary repertoire, and the role of Aboriginal expertise in its interpretation, to reinforce the High Court's tentative incorporation of indigenous laws and customs into Australian common law. However, the legalistic vocabulary of *'native* title' which both adopt and the lingering purchase of race as a categorical marker of Aboriginality nourished more practised prejudices invested in the term 'native' and in those it designates, with which this chapter began (see Pearson, 1993c: 78). In the careful wording of the Act, and the Court, Aboriginal genealogy of itself does not constitute evidence of native title. The primary criteria of 'a continuous connection to the land' placed the onus of proof on the performative continuity or practical observance of traditional laws and customs pre-dating 'settlement' by community members today, however fragmented or dispersed they had become (Keon-Cohen, 1993).[22] As the Labor Party Leader in the Senate explained in rejecting a proposed Green Party amendment to these provisions in the Bill,

> we see a continuing connection of some kind, not necessarily physical, with the area in question in accordance with their customs and laws still needing to be established. . . . Where, for example, a particular sacred site is on the land in question . . . that site may be inaccessible [due to fencing, for example], but it still has dreaming stories about it which form part of the continuing character of the particular community and while that community may be physically detached from the sacred site, that site is still very much a part of the community's ongoing cultural environment. (Senator Gareth Evans, Hansard (Senate), 16/12/93: 5347 and 5349)

Cross-party acknowledgement that only a fraction of the already small number of indigenous Australians were likely to benefit from native title under the provisions of the Act arose in large part from this distinction between genealogy and cultural practice.

The legal significance of the term 'native title', as a fragile but continuous thread of recognition that the indigenous peoples of colonized territories observed their own proprietory laws and customs, was easily corrupted. Undaunted by the climate of 'political correctness' or the '7 per cent' estimate of potential beneficiaries, some Opposition speakers persisted in framing the legislation in racial terms. Peter Slipper (Queensland, Liberal Party) was not alone in declaring it 'racist', amounting to 'apartheid in reverse' by giving

> . . . one group of Australians virtually a blank cheque at the expense of other Australians. In other words, all Australians are equal, but some are more equal than others. (Hansard (HoR), 23/11/93: 3475)

But others of his Coalition colleagues were more concerned by the porosity of the racial categories sustaining this perverse analogy, and which could open native title up to all manner of 'Aboriginal pretenders'. For Senator Panizza (Western Australia, Liberal Party), for example, any 'dilution' of the supposed geneological parameters of 'Aboriginality' threatened to swell the ranks of claimants alarmingly if left unchecked.

> As the years go by what will happen to the degree of Aboriginality? . . . in the racing game, the cattle industry or anything like that, entries are out of the stud book after four generations. I want to know what will be the minimum claim for Aboriginality in 99 years time? (Hansard (Senate), 16/12/93: 5473)[23]

While for some, like Raymond Braithwaite (Queensland, National Party), it was the possibility that Aboriginality might *not* be confined to a matter of genealogy that seemed to be most troubling:

> A lot of people want to know what Aboriginality is. . . . As I understand it, if I were accepted by an Aboriginal community and were prepared to accept Aboriginal ways and live with that Aboriginal community, I could be classified as an Aboriginal. (Hansard (HoR), 25/11/93: 3751)

Such nostalgic urges to cement the place of the 'native' in the Australian body politic by evoking the certainties of a racially fortified categorization of Aboriginality expose their own treacherousness. They also mark the discursive limits of this Parliament, attracting cross-party opprobrium and locating 'the R word' at the extremities of the Opposition benches. But the malignance of such obdurate racism constantly overspills these convenient confines in the pages of Hansard, lingering in a widely held conviction that 'the' Aboriginal relationship to land was 'too unsophisticated' to involve any form of ownership. It is a conviction that ripples through the 'commonsense' currency of conservative Australian opinion in which the familiar itch of the 'primitive' and the universal pretensions of a Eurocentric definition of property is given a new twist, characterized by Gelder and Jacobs as the 'new racism' (1998). The political and legal encoding of property in Europe, as we saw in the previous section, privileges settled cultivation. This privileging itself presumes and reinforces a conceptual separation between the realms of society and nature, and their mutual reordering through the vortex of labour. Mapped back into and through the racialization of 'Aborigines', this proprietorial assemblage is as complicit in displacing them from the company of 'civilized men' into that of de Vattel's 'roving beasts' as any scientific enterprise (see Vogel, 1988).[24] As the maveric former Liberal Party Senator, Fred Chaney, remarked of his fellow Western Australians in a series of media interviews during the Native Title debates,

> The idea that Aboriginals have some rights of their own, as against privileges which are given to them by a benevolent government, is quite foreign to Western Australian thinking. . . . An attitude [remains] that

Aboriginals can almost be treated like kangaroos and simply moved on whenever there is a competing (land) use. (quoted by Chris Haviland (NSW, Labor Party), Hansard (HoR), 23/11/93: 3484)

The self-fulfilling and long-buried premise of this posture, for all its twists and turns, is one of the most resilient threads connecting Colonial and Commonwealth parliamentary deliberations. Just as its traces are exposed by those opposing the repudiation of this premise in the Mabo ruling and Native Title legislation in the Australian Parliament today, so the record of voices against the grain of the colonial Acts of the nineteenth-century British Parliament complicate its historical career. The salience of property to the racialization of Aboriginal peoples is illustrated with unselfconscious clarity by a sequence of questions and answers in the evidence of Archdeacon Broughton to the British House of Commons Select Committee in 1835 on the treatment of Aborigines in New South Wales.

Q: Have they any property, properly so-called?
A: No property; they wear a portion of clothing which those in the interior can scarcely be said to do.
Q: Nothing like assignments of land have been made to them?
A: They were made by Governor Macquarie, but it was found impossible to attach them to the soil.
Q: They were totally averse to cultivating then?
A: Quite averse to permanent labour of any sort. I should say, that they have a notion among themselves of certain portions of the country belonging to their own particular tribe; they have frequently said to me that such a part was their property; but that is all assigned now to Europeans. (BPP, 3/8/1835: paras 232–234)

In the same vein various Opposition speakers in the Australian parliament sought to confine the purchase of the Mabo ruling on Native Title to the Murray Islands on the grounds that the Meriam people who brought the case were 'gardeners' who cultivated family plots, whereas 'mainland Aborigines' had no such appreciable traditions.[25]

However, this apparent continuity of reasoning belies significant discrepancies in the ways it was pursued and the conclusions drawn by these distant sets of parliamentarians. In the Australian Native Title debate, numerous Opposition speakers confidently recite as a matter of settled historical fact that

... [Australia's] native inhabitants did not intensely utilise their land, as much as simply traversing it, living as a nomadic people. ... The reason that no treaties were made does not represent a denial of Aboriginal existence but, rather, that there was no leadership or cultural structure with which to negotiate. (Tim Fischer (Leader of the National Party), (Hansard (HoR), 23/11/93: 3425)

The equally incontrovertible corollary, they conclude, was the legal incorporation of this 'fact' in the doctrine of *terra nullius* which legitimated

> the acquisition of any territory inhabited by peoples whose civilisation was thought to be less developed in terms of ownership of property or political organisation. (Senator Hill (Senate Leader of the Opposition), Hansard (Senate), 14/12/93: 4578)

The Report of the British Parliamentary Select Committee over 150 years earlier covers the same ground but to very different effect.

> Such indeed is the barbarous state of these people, and so entirely destitute are they even of the rudest forms of civil polity, that their claims, whether as sovereigns or proprietors of the soil, have been utterly disregarded. The land has been taken from them without the assertion of any other title than that of superior force and by the commissions under which the Australian colonies are governed. (BPP, 1837: 84)

This account of the contemporary colonial perception of the political organization of Aboriginal society is familiar but its import resides in the settlers' unjustifiable response to the situation. Far from constituting legitimate grounds for their dispossession, the Committee argued that

> So far as the lands of the aborigines are within any territory over which the dominium of the Crown extends the acquisition of them by Her Majesty's subjects, upon any title of purchase, grant or otherwise, from their present proprietors should be declared null and void. (BPP, 1837: 78)

It was a conclusion more in tune with the Mabo ruling than with the comfortable histories of its parliamentary critics, and which identified a remedial course of action more radical than anything in the Native Title legislation.

As this disconcerting conjunction of racisms suggests, the displacement of Aboriginal peoples in the Australian polity who, by definition, were 'there first' is configured in and through their relations with the categorical figure of the 'settler' which harnesses the land to a homogeneous territorial imaginary and, thereby, occupies the space of the nation. This imaginary has been nourished for most of this century by a white Australia immigration policy, pursued by successive Commonwealth Governments until the Whitlam Labor administration dismantled it in 1972 (Markus, 1994). The National and Liberal Parties perpetuate this territorialization of the 'settler' today by representing pastoralists and others who work the land as the embodiment of Australian national identity. John Hewson, Leader of the Liberal Party and key opponent of the Labor Government's Native Title legislation in the early 1990s, accused the then Prime Minister of 'selling out the interests of average Australians, he is selling out the miners, he is selling out the farmers and he is selling out those who want jobs' (Hansard (HoR), 25/11/93: 3742). Here the statistical polity of the 'average Australian' is neatly equated with such 'men of the soil'. But for many of his fellow Liberals, like Barry Wakelin (South Australia), these same people had been rudely sidelined to the social and geographical margins of the country: 'As regional Australians we are stereotyped and denigrated by many in the media and many of our political opponents' (Hansard (HoR), 25/11/93: 3744). His fears are fuelled by those on the Government benches and in the national press (see figure 4.4), who caricatured the public appeal of Liberal and National Party rhetoric in terms of 'the redneck conservative elements in our society' (Garrie Gibson (Queensland, Labor Party), Hansard (HoR), 23/11/1993: 3421) and 'the ultra conservative rump of our community' (Hon. Francis Walker (NSW, Labor Party), Hansard (HoR), 25/11/93: 3734).

This precarious positioning of pastoralists/settlers in the assemblage of Australia is not, as their representatives imply, a novel one. As at least one

Figure 4.4 'Rednecks': the pastoralist as metropolitan caricature (cartoonist: Shakespeare; *Sydney Morning Herald*, 11/11/93)

parliamentarian reminds the House, their status as political subjects has been ambivalent, at best, from the outset of colonial government.

> Much of this land is owned by the descendants of those early squatters who gained this land at no monetary cost. We now often revere these people as pioneers who were willing to go into the outback, fight the hostile natives and gain this land for themselves. (Maggie Deahm (NSW, Labor Party), Hansard (HoR), 23/11/93: 3473)

The British Parliamentary Committee on Aborigines (1836–37), to which she goes on to refer, were rather less reverent, attributing what it saw as the 'disastrous oppression of natives' in the colonies to

> . . . ignorance [and] the difficulty which distance interposes in checking the cupidity and punishing the crimes of that adventurous class of Europeans who lead the way in penetrating the territory of uncivilized man. . . . This then appears to be the moment for the nation to declare that with all its desire to give encouragement to emigration and to find a soil to which our surplus population may retreat, it will tolerate no scheme which implies violence or fraud in taking possession of such a territory. (BPP, 1837: 75–6)

It went on to conclude that 'The protection of aborigines should be the duty of the Executive government, administered in Great Britain or by Governors of Colonies not by local legislatures . . . made up of settlers who have a vested interest in undermining native rights' (BPP, 1837: 78).[26]

All the more startling then that in the face of their late twentieth century predicament as a self-ascribed minority in the national polity, their parliamentary representatives should align themselves with Aboriginal people as co-habitants of the 'outback' and, thereby, as custodians of the 'real' Australia, even as they continued to prosecute their 'vested interest in undermining native rights'. Barry Wakelin's concern with the denigration of pastoralists as 'regional Australians', for example, lead him to claim that

> we have been the closest to the aboriginal people for many generations and we share in many ways some understanding of Australia's urban dominant culture. The comfortable clichés of the ignorant and the politically correct roll off the tongue as many confess that they have had little or no contact with aboriginal people. (Hansard (HoR), 25/11/93: 3744)

Tim Fischer, Leader of the National Party, was even more emphatic:

> The National Party, above all, understands what it means to have a strong attachment to the land. There are families in Australia who have held their land for generations. Their forebears worked the land and were

> buried on the land. . . . It is not correct in all the circumstances to argue
> that this attachment to the land is qualitatively different from the deep
> attachment felt by Aboriginals and that the land has greater significance
> to many native tribes. (ibid., 23/11/93: 3427)

The pastoralist emerging here from the sober pages of Hansard bears a
passing resemblance to that all-Australian Hollywood hero Crocodile
Dundee, allying Aboriginal bushlore and Anglo-Saxon certitude to cut a
swathe through the enervating sophistications and rampant vices of city life
(see Morris, 1988). As a political gambit, rallying the nation around such a
figure signals a desperate crisis of identity and a regional rift in the body
politic as potent as that between indigenous and non-indigenous Australians. Opposition politicians put themselves in the untenable position of
deriding the 'land ignorant' urban majority as 'tourist[s] from the soft areas
in Australia' (Senator John Panizza (Western Australia, Liberal Party),
Hansard (Senate), 16/12/93: 5413), while appealing to them as fellow
proprietors by insinuating that native title threatened

> . . . thousands upon thousands of other Australians with all sorts of
> interests, as well as normal other Australians who have a quarter acre
> block in the back blocks of capital cities who worry about it as well.
> (Senator Ian Campbell (WA, Liberal Party), Hansard (Senate), 17/12/93:
> 5011).[27]

Far from consolidating the space of the nation, like territorial tarmac, the
settler designation takes and loses shape in a mass of cross-cutting social
currents that do not originate in, or arrrive at, the same place in their
journeys towards becoming Australian (see Rowse, 1993c).

Australia's tortuous negotiation of the etymological ties between
'native' and 'nation', which nourish European political imaginaries and
nationalist mythologies more widely, cuts to the quick of a persistent
dilemma for the settler-nation. The hyphen which holds this national
identity in place has become ever more tenuous as the pastoralist-hero who
embodies it is figuratively and socially displaced and the (re)assemblage of
Aboriginality as a practical polity emerges from its shadow. The categorical
'we' of the 'settler' appears from these debates in no less disarray than the
categorical 'other' of the 'native'. For all its pliablity, the 'average Australian' being conjured in the parliamentary chamber and record, and circulating the globe in the celluloid community of *Neighbours*, does not inhabit
this territorial imaginary. But neither does such a polity of averages respect
difference or countenance its expression in the fabric of governance in the
ways that both the Mabo judgment and Native Title Act tentatively begin
to do.

becoming a post-colonial country

> It needs to be recognised that treating unequals equally can infringe the principle of equality before the law as much as by treating equals unequally. (Report of the Royal Commission into Aboriginal Deaths in Custody, 1991: 5)

The Mabo ruling fuelled public anxieties in Australia which exceeded the settler/Aboriginal coordinates of the disputed space of the nation. Forceful political realignments in their name certainly took shape. Statutory bodies like the ATSIC were brought into the Commonwealth's Native Title policy-making process in unprecedented ways even if the tribunals set in place by the Act have not lived up to its promise in practice (http//www.native.title.gov.aus). The highly organized mining and pastoral lobbies orchestrated opposition to this legislative response through concerted media campaigns (see Gelder and Jacobs, 1998). And a new party, 'One Nation', exploited the reactionary political space it opened up in State and Commonwealth elections. But for most Australians living in cities and towns around the coast, including some 40 per cent of indigenous Australians, situating themselves in this historical and territorial renegotiation proved a much more complicated business.

Mabo, and the Native Title legislation which it inspired, evokes continuities and tensions in the contested and restless constitution of the Australian body politic. They provide pertinent reminders that territorial struggles are not about a zero-sum allocation of a finite area of land but articulate the pivotal role of property and sovereignty in constituting persons and things, states and citizens and stablizing their boundaries and relations in the socio-material ordering of the country. These territorial practices are among the most powerful and durable parameters of governance, inhering in the categorical configuration of 'settlers' and 'natives' as much as the wire fences that demarcate the land. Indeed, as reactionary responses to the admission of 'native title' illustrate, they are the bedrock of Australian histories and geographies that have effaced all trace of their fraught and patchy making and become as definitive of this island continent as its ocean shoreline. However, as I have suggested in this chapter, even these most intractable parameters of territorial governance are in practice plastic achievements. The legal and parliamentary documents which freight their purchase in space–time are themselves uncertain messengers in the intricate assemblage of the law of the land. To cast territoriality in performative terms is not to diminish its potency or durability but to focus attention on the tangle of socio-material agents and frictional alignments in which it is suspended and to recognize that they harbour other possibilities.

The Mabo judgment represents just such a significant realignment of the parameters of settlement. Rather than coming out of the blue, the majority judges mobilized fragile but persistent currents in English common law through the wormholes of territorial governance that admitted 'native title' and which had been woven into the constitutional fabric of other settler-nations already. The most significant of these wormholes is the epistemology of the common law itself which, as Tim Murphy suggests, observes a medieval textuality that is 'utterly indifferent to the pastness of the past'; as much a manner of speaking in which the pivotal moment is always now as a written code that foreshadows its interpretation (1994: 77). In the same breath, the Mabo judgment voiced an unwelcome reminder to some of Australia's largest landed interests that the leases and licences which have sustained their own claims on the land, and those of their forebears, fell well short of the heroic pretensions to unqualified possession that they had accustomed themselves to assume.

On the narrowest interpretation, the Mabo judgment and Native Title Act 1993 acknowledged the tenacious continuity of Aboriginal attachments to the land in the interstices of 'settlement', a tenacity confined to a minority of indigenous Australians and to some of the remotest parts of the country. The larger significance of these documentary moments lies in their cautious recognition of difference and multiplicity in the fabric of governance, substantive markers in Australia's protracted shift towards becoming a post-colonial nation. The incorporation of native title into the body of the law departed from a political framing of Aboriginal interests in terms of welfare delivery to a 'disadvantaged race'. Instead, it was a modest and belated step towards recognizing the Aboriginal presence in the nation as a practical polity whose proprietary customs constituted legitimate rights on their own terms to a stake in the land and the distribution of benefits accruing from it, removing the 'settlement' fiction of *terra nullius* (see figure 4.5). Just as importantly, both the judgment and the Act complicated the spaces of dis/possession by countenancing the co-existence of native and other forms of title to the same area of land, an acknowledgement of 'proprietary pluralism' reaffirmed some three years later by another High Court ruling (*Wik* v. *Queensland*, CLR, 1996).

The register of 'reconciliation' which these documentary moments share was bound up with a wider political climate of national transition. The Keating Government allied its fortunes to a realignment of Australia's civic identity and institutions to accommodate the manifest cultural diversity that constitutes its hybrid social texture today. As James Tully notes of another nation navigating a passage from a 'settler' to a 'post-colonial' identity, Canada, such a transition involves

> more than a civic awareness that citizens of other cultures exist in one's polity. One's identity as a citizen is inseparable from a shared history with

Figure 4.5 *Terra nullius* II: pay the rent (*Aboriginal Law Bulletin*, 1992, 2/57:5). (Photo by: Sandy Scheltema, courtesy *The Age*)

other citizens who are irreducibly different; whose cultures have inter-acted with and enriched one's own and made their mark on the basic institutions of society. (1995: 205)

Such a reassemblage of the political fabric of the nation is anything but straightforward. Those parliamentary parties which had opposed the Native Title Act 1993 became the next Government and have set about re-asserting the territoriality of 'settlement' through the political rhetoric of 'One Australia'. As Leader of Opposition Business under the Keating administration, the new Prime Minister John Howard (Liberal Party), like many of his Coalition colleagues, had rejected native title as a threat to the integrity of the settler-nation – 'one country, one law, all of us living under the same rules'. His administration's response to *Wik* could not have been more in contrast to its predecessor's response to Mabo. Under pressure from the pastoral lobby to strengthen their tenurial grip by converting pastoral leases into freehold title, the Howard Government passed an Act of Amendment watering down the provisions of the Native Title Act (see Parliament of the Commonwealth of Australia, 1998).[28]

But post-Mabo, the tired liberal appeal to formal equality which assimilates everyone into 'one-nation' by treating them 'the same' will no longer wash. Its treachery is epitomized in Howard's recent refusal to 'apologize' for the policy of 'assimiliation' (26/8/99) that forcibly removed

Aboriginal children from their families, and which he sought instead to consign to the past as 'the most blemished chapter in our national history' (see Poole, 2000; Frow, 2001). Like land rights, this is not a chapter that can be closed but an unsettling pulse in the body politic of a country struggling towards a meaningful territorial renegotiation and accommodation of difference in the space of the nation.

Reinventing Possession:

boundary disputes in the governance of plant genetic resources

> On the horizon are a whole new set of claims to proprietorship. . . .
> They arise *out of* the very perception of hybrids, out of mixes of
> techniques and persons, out of combinations of the human and non-
> human, out of the interdigitation of different cultural practices. Not
> socially innocent, not without their own likely effects, they presage
> new projects of modernity. (Marilyn Strathern, 1999b: 122)

geo-politics in/of the flesh

On the busy calendar of international campaigns promoted by the United
Nations (UN), 16 October is designated as 'World Food Day' to mark the
anniversary of the founding of the Food and Agriculture Organization
(FAO). The stated ambition of the FAO is to secure 'food for all' as a
human right, and in 1993 it dedicated World Food Day to making
connections between this mandate and sustaining 'the biological diversity
of our planet'. The messenger of this event was a glossy document entitled
Harvesting nature's diversity (FAO, 1993), spreading the word in each of
the five official languages of the UN. In the foreword to this document, the
Director-General explains its premise thus.

> Humanity's place in nature is still not widely understood. Human
> influences on the environment are all pervasive: even those ecosystems
> that appear most 'natural' have been altered directly or indirectly during
> the course of time. Starting some 12,000 years ago, our forebears, as
> farmers, fishermen, hunters and foresters, have created a rich diversity of
> productive ecosystems. . . . Once lost, this heritage cannot be recovered
> or restored. (FAO, 1993: 1)

At first sight, this may seem an unremarkable statement of the obvious. Its
heretical potency only snags our attention when set against the louder

claims reverberating in another chamber in the UN labyrinth – its Environment Programme. The Convention on Biological Diversity (CBD) (1992),[1] a document which has the force of law, commits signatory states to

> the conservation of biological diversity, the sustainable use of its components and the fair and equitable sharing of the benefits arising out of the utilization of genetic resources, including by appropriate access to genetic resources and by appropriate transfer of relevant technologies. (CBD, 1992: Article 1)

Here, 'diversity' is defined as 'the variability among living organisms from all sources . . .; this includes diversity within species, between species and of ecosystems' (ibid.: Article 2).

Where the FAO version articulates diversity as a heterogeneous achievement in which human being and doing is enmeshed through long and situated association in the spatial and corporeal fabric of botanical becomings, the CBD account casts it in wholly biological terms, the outcome of an evolutionary process divested of human presence. The one conjures a world that is hybrid 'all the way down', enfolding humanity in its ceaseless commotion time out of mind. The other conjures a world until recently unmarked by the (invariably negative) 'impacts' of human society, only countenancing hybridity as a technological accomplishment associated with the advent of 'genetic resources'. The intensely weighed and closely vetted vocabularies of these international policy documents anticipate Strathern's assertion that the 'very perception of hybrids' is generating a 'whole new set of claims to proprietorship' by several years. More importantly, they bear witness to the political charge of hybridity in the fraught assemblage of 'nature's diversity' as the latest in a catalogue of phenomena to be (re)configured as a terrain of global environmental governance. Tagging the convoluted careers of such documents as both the artefacts and mediators of geo-political struggles, affords us glimpses of the laborious practices and serious stakes involved in 'making the cut' that demarcates the natural from the social in the flesh of plants and (re)aligns its germinal potencies in the fabric of human attachments. The terms on which hybridity is being codified in legal protocols governing rights in, and jurisdictions over, plant germplasm heralds a de/re-territorizalization of the vital associations between plants and people no less consequential than that effected through the calculus of cultivation in the lands enclosed by European colonialism.

This contrapuntal staging of the very different configurations of 'nature's diversity' articulated in the FAO and CBD documents is more than just an analytical contrivance. It amplifies ongoing disputes between shifting alliances of nation states and rival organizational networks within the institutional apparatus of the UN to render the socio-material fabric of plants governable by subjecting its lively rhythms and patterns to the orderly disciplines of the law. As Donna Haraway suggests:

> The question of [what] kind of materiality genes are going to have for different sorts of communities in the world is absolutely on the table, it's molten. . . . Multiple constituencies are daily engaged these days in . . . these sorts of . . . translations . . . rarely [as a result of] goodwill and choice, but literally being forced into some kind of exchange relationship where genes are the boundary objects. (1995: 517)

To be effective, boundary objects have to be sufficiently plastic to satisfy the informational requirements of several communities of practice yet robust enough to bridge differences and so sustain working arrangements, without imposing a naturalization of categories (see Bowker and Star, 1999: 297). Plant Genetic Resources (PGR) have just these qualities, diagramming between communities, states and regions rich in plant genetic diversity, notably Africa, Asia and Latin America, and those, notably North America and Western Europe, generating the socio-technical and legal means to exploit it commercially (see Kloppenburg, 1988; Fowler and Mooney, 1990). This burgeoning traffic (re)convenes struggles between 'first' and 'third' world nations, multinational corporations and peasant farmers, pitting patent law against plant lore in the intimate fabric of knowledge practices that body forth in plant usage. Through the campaigning activities of non-governmental organizations (see Posey and Dutfield, 1996: 245–81) PGR have become familiar envoys of global injustice in the virtual political landscapes of the mass media as, for example, in the broadsheet advertisement by the UK charity ActionAid illustrated in figure 5.1.[2]

In this chapter I want to dwell on a particular moment in the shifting topology of the global governance of PGR, in which these collective acts of translation can be observed in the (un)making – the fraught assemblage of an International Undertaking on Plant Genetic Resources under the auspices of the FAO in the 1980s (see Cooper, 1993). As the long-serving Secretary of the Commission established to prosecute this Undertaking (CPGR)[3] recalled in an interview, it plied uncertain passage in already hostile waters:

> . . . An old colleague of mine headed the Spanish delegation and was elected conference chairman [*sic*] in 1979. We discussed the PGR issue and [he] agreed to present an argument for a network of genebanks under FAO auspices through the Spanish delegation. [He] asked me to prepare a draft resolution on a regulatory framework (which I did in consultation with FAO legal advisors). Under pressure from the US delegation, we finally did not submit it to conference, but the argument

Figure 5.1 ActionAid poster campaign: 'This will make you sick' (*The Observer*, 2/7/00)

was echoed by other delegates, particularly from India. In 1981, it was taken up much more forcefully by the Group of 77, led by the Mexican delegation who introduced the draft resolution. (interview, 7/7/94)[4]

As stipulated in Resolution 8/83 of the 22nd session of the FAO conference in November 1983,

> The objective of this Undertaking is to ensure the PGR of economic and/ or social interest, particularly for agriculture, will be explored, preserved, evaluated and made available for plant breeding and scientific purposes. This Undertaking is based on the universally accepted principle that PGR are a heritage of mankind and consequently should be available without restriction. (Article 1, FAO, 1983a: 50)

Despite, or perhaps, because of, its modest legal force, this non-binding agreement gathered political momentum, accumulating additional powers and instruments through a series of subsequent resolutions that came to assume the collective habit of a 'Global System for the Conservation and Utilization of Plant Genetic Resources' (CPGR, 1993).

Ten years on, over 130 countries could be counted as adherents to the International Undertaking or members of the Commission for Plant Genetic Resources. At its fifth session in 1993 the Commission secretariat elaborated the organizational fabric of this 'Global System' in/as a diagram (reproduced in figure 5.2), aware that it was as much a mapping of what could be, as of what had been, accomplished.

In practice, this orderly contrivance belied a brittle alignment of forces spun between national governments and FAO bureaucrats; delegates and documents; legal protocols and technical procedures; talking and voting. Not only did the portable inscriptions of collective resolution (C/8/83, C/4/89, etc.) fail to paper over the cracks of persistent dissent, notably by the United States,[5] but the graphic coordinates of this 'global system' had been disconcerted by the very different political impetus and instruments of the Convention on Biological Diversity even before the ink was dry. At the same time, its harnessing of political and legal energies to the constitution of PGR as a 'heritage of mankind' marks a more durable interference in the spatial practices of global environmental governance, testing the territorial vernacular of sovereignty and property to their limits (Litfin, 1998).

I have chosen this moment in the assemblage of PGR precisely because its precarious purchase refuses the analytical urge to solidify it as either the foundation or culmination of some evolutionary process of global governance. Rather than a point of departure or arrival, the International Undertaking emerges as a mediator in a provisional mode of ordering that invites another way of travelling through the commotion of currents caught up in the business of rendering PGR governable. Against the linear habits of an institutional biography of CPGR or a developmental account of the International Undertaking (see, for example, Kloppenburg and Kleinman, 1987a), the (de)composition of these happenings charts a mode of ordering that is neither a discrete instance nor a complete state, but rather an immanent gathering of forces.[6] My journeying through the collective

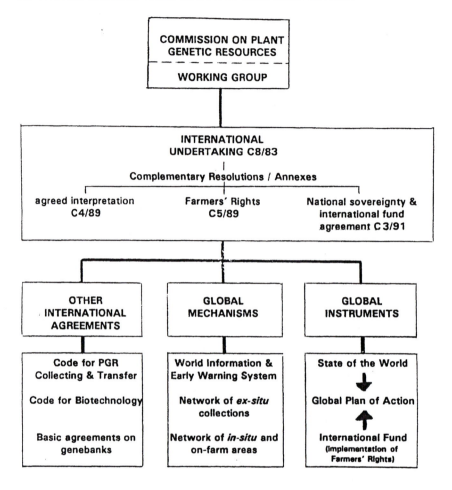

Figure 5.2 Governing PGR: 'the Global system' (CPGR, 1993: agenda item 4, p. 14)

practices of this event mimics the assemblage of PGR, a coming and going between talk and text, laboratories and fields, plants and people, knowledges and rights, that exhausts neither the promiscuous bio-geographies of plant association nor the geo-political possibilities of governing them. It also bears witness to my own navigation of the FAO labyrinth and its documentary archives, including verbatim records, legal instruments, minuted agreements and expert reports, in which the governmentality of PGR inheres.[7] Pursuing this signal acronym through these paper practices affords glimpses of this boundary object at work, which are at once pale echoes of the collective talk that it convenes and true to the lexical conduct of law-making that intensifies its affects (see Goodrich *et al.*, 1994).

In tracing constitutional tensions and fault-lines in the assemblage of the International Undertaking, I want to hold on to their creative liminality, not as divisions which prefigure this event, but rather as cracks inhabited by its collective practices – in-between spaces that afford possibilities for becoming otherwise. It is here that we can observe boundaries in the (un)making as new proprietorial configurations of persons and things are cut out of/into the already segmented fabric of human/plant associations. My purpose is to interrogate these labours of division in terms of the legal encoding of three such 'cuts' in the Resolutions of this governing collective and the political disputes that interfere in their would be de/re-territorializations of PGR. The first is the cut between nature and culture in the space of the flesh, the line that marks the evidentiary boundary of legal claims to ownership, enrolling genes as material witnesses to the discernment of such claims in the fabric of plants themselves (see Battaglia, 1994). Here, the disputed question of what constitutes PGR in the terms of the Undertaking comes under scrutiny. The second is the cut between global and national in the space of the polity, the line that configures the 'we' in whose name the collective governs, carving out its jurisdiction over PGR as a 'heritage of humankind' from the practised territorialities of sovereign nation states (see Lipschutz, 1998). Here the issue of where political authority over PGR should reside comes into focus. The third is the cut between heritage and invention in the space of the subject, the line that constitutes singular creative acts/agents from the spatio-temporal flux of botanical knowledge practices, prescribing the person(s) to whom such innovations are attributable and entitling them to proprietary benefits in their future use (see Strathern, 1998). Here, controversies about what/whose practices count as knowledge production/ers take centre stage.

1st cut – between nature and culture

> Probably, the total genetic change achieved by farmers over the millennia was far greater than that achieved by the last hundred or two years of more systematic science-based effort. (Simmonds, 1979: 11)

The first requirement of any governance regime is to define its object, an activity that is more fraught than this familiar formulation makes it appear. In the case of the International Undertaking, elaborating an answer to the question of what constitutes PGR took five years of consultation, expert opinion and debate, not to mention reams of paper, and still settled little. The mutable potencies of plants and their lively habits of association with soils and insects, people and water, proved no easier to circumscribe as 'genetic' entities than they had as 'varietal' forms in earlier efforts under UPOV to establish plant breeders' rights (see Berlan and Lewontin, 1986: 787). Even the formal definition of PGR eventually codified in the legal

framework of the International Undertaking betrays the messy complications of drawing a line between nature and culture and of its holding steady
in the objects that are enjoined to bear witness to this boundary.[8]

> For the purposes of this Undertaking, PGR means the reproductive or
> vegetative propagating material of the following categories of plants:
>   i. cultivated varieties (cultivars) in current use and newly developed
>   varieties;
>   ii. obsolete cultivars;
>   iii. primitive cultivars (land races);
>   iv. wild and weed species, near relatives of cultivated varieties;
>   v. special genetic stocks (including elite and current breeders' lines and
>   mutants. (Article 2(a) FAO, 1983a: 50)

As the disconcerting combination of precision and indeterminacy in
this catalogue of distinctions makes clear, nature and culture do not divulge
themselves in the fabric of plants like some sort of botanical apartheid that
marks out the wild and the domesticated as certain kinds. But neither are
they merely the projection of human categories on to an object that makes
no difference to their effectivity. Rather, in the manner of Rheinberger's
(1997) epistemic things or Latour's 'factishes' (1999a), PGR emerge as a
socio-material fabrication in which the histories and geographies of more
than vegetative associations that they make flesh are constituted through
and constitutive of this ordering event. In the multi-lingual business of the
FAO Conference, PGR proved a ready envoy of collective unanimity in
terms of its vital importance to world agriculture and food security, even as
national delegates disputed just what 'it' was. Such disputes enjoin different
understandings of the fusion of human and non-human energies embodied
in cultivated plants, freighting their hybridity with political consequences
for the determination of sovereignty and property. For the Mexican
delegate introducing the draft resolution on PGR to the FAO Conference in
1983, and speaking for the Group of 77 coalition of developing countries
(see note 4 above), the fabric of associations between plants and people is
so densely woven as to render their disentanglement perverse.

> The labour of generations of peasants around the world, as well as the
> gifts of nature, has allowed us to develop a long history of riches as
> indispensable to our survival as the air we breathe. (Lopez-Portillo, FAO,
> 1983b: 294, author's translation from Spanish)

From this perspective, hybridity is a mode of worldly inhabitation that
precedes the urge to separate out the social from the natural rather than a
gesture towards their reconciliation (see Ingold, 1993). Here, 'making the
cut' is a political exercise which has been conducted through western legal
and scientific practices that look to themselves as the benchmark for
substantiating social attributes in plant germplasm, disqualifying other

modes of association and fostering long habits of unequal exchange between north and south. In this spirit, proponents of the Undertaking saw it as an opportunity for those at the sharp end of such exchanges to challenge the legitimacy of this benchmark and expose its adverse consequences not only for those that it dispossessed but also for the erosion of plant biodiversity. At the next Conference two years later, Lopez-Portillo warmed to his theme.

> Historically, all manner of civilisations have depended on, or created, hundreds of thousands of plant species and varieties for their daily sustenance, for health and hygiene, for clothing, shelter, for light and energy, for obtaining dyes and chemicals, for their magic and religion, for their industries and wars, for their symbols and progress. In sum, for the harmony and stability of their geography, their society and their culture. . . . However, the western concept of civilisation and development and the far-reaching economic and cultural subordination implicated in the modernization of agriculture and eating habits, over several centuries, has resulted in the significant reduction in the number of plant species and cultivated varieties we see today. . . . (Lopez-Portillo, FAO, 1985a: 293, author's translation from Spanish)

For opponents of the International Undertaking, principally those speaking for countries at the forefront of instituting this benchmark and the loci of its main economic beneficiaries, this argument and the Resolution it informs marked the politicization of what to date had been a technical issue objectively conducted by 'science'.

The Canadian delegate, for example, sought to diminish the confusing hybridity of PGR by inflating the compass of plant breeding from cultivated species and their near relatives to the totality of plant life, so reaffirming the contribution of a nature untainted by social claims. In the same breath, he minimizes the relative significance of scientific interventions in the (small) social contribution to plant breeding, obscuring their critical association with its commercialization and distributive consequences. Adopting the self-effacing facticity of science, he argued that:

> Much of the misunderstanding which led to the politicization of this issue comes from an abundance of opinions and a scarcity of factual information on plant breeding. Fact number one. Nature accounts for 90% of all plant breeding to this day and most of the balance is the result of movement and manipulation of germplasm by millions of individuals from countless cultures for thousands of years. The work done in research laboratories during the last two centuries accounts for far less than 1% of all plant breeding. (Fredette, FAO, 1985a: 298)

Such false modesty is shortlived. In the next sentence he enjoins science as the guarantor of this 'factual' division of labours, retracing its own effects in/as the corporeal cartography of plant germplasm, noting that:

> The term germplasm was only coined by German scientist August
> Weismann in 1883, and the first attempt to deliberately search for,
> classify and preserve wild plants and primitive cultivars from around the
> world goes back to the pioneering work of Nikolai Vavilov in the 1920s.
> (Fredette, FAO, 1985a: 299)[9]

From this perspective, hybridity is a scientific achievement that brings
social order to an a priori world of nature. But just as the solidification of
prejudicial practices into legal certainties proved an unreliable basis for
interpreting the constitutional territoriality of Australian settlement, so too
was this coagulation of knowledge practices into hard facts to prove a less
robust ally than opponents to the Undertaking would have it.

At the time of the Undertaking, 'genetic fingerprinting' techniques
were affording new and more intimate mappings of the heritable material
of plants and plant populations. But as the authors of a background study
paper prepared at the request of the CPGR secretariat for its first extraordinary session in 1994 were to suggest, these techniques raised as many
questions as they answered for the discernment of discrete or stable genetic
entities. They concluded that:

> It is impossible, even at the present state of increased knowledge, to
> earmark genetic entities 'beyond reasonable doubt' to any specific genetic
> source-materials. In fact it is doubtful whether it will ever be possible.
> Genetic diversity results from random events of mutation that may occur
> at any time and in any population. Through selection the frequency of
> specific genes or gene-combinations may be increased or decreased, but
> not its actual occurrence. Uniqueness as a principle therefore makes in
> fact no biological sense. (Hardon et al., 1994: 16)

As well as a mapping in the flesh, these genetic techniques were also being
used to retrace earlier scientific cartographies of plant germplasm in terms
of its global distribution. In other words, the question of what constitutes
PGR was closely bound up with fixing its location. Here the space of the
flesh and the business of demarcating the wild from the cultivated are
further complicated by the complex motions of plants and people and
multiple foldings of time–spaces that mark their passage. These complications in the natural histories and bio-geographies of plants are now well
documented.[10] The collective mapping of PGR against this shifting topology of plant associations afforded opportunities for new practices of
division, aligning 'wild' varieties and 'land-races' with native habitats or *in-
situ* sites and 'cultivars' and 'breeders' lines' with scientific collections or
ex-situ sites. This bi-partite siting of plant germplasm and the governance
practices which it freights mirrored and entrenched the geo-political alignments of countries promoting and opposing the International Undertaking,

placing the legal status and governmentality of this territorialization in constant dispute (Kloppenburg and Kleinman, 1987b).

For the purposes of the Undertaking *in-situ* sites refer to 'locations of plant genetic resources in their natural habitat' (Friis-Hansen, 1994), primarily associated with those regions and countries designated as the 'third world'. This collective cartography of PGR explicitly followed Vavilov's phytogeography of cultivated plants, a debt acknowledged in the text accompanying the glossy version published in *Harvesting nature's diversity* (FAO, 1993) which is reproduced in figure 5.3.[11] The difficulty for those, like the Canadian delegate, who would rely on this mapping as a secure scientific foundation is that its classificatory predicates, the concepts of centres of agricultural origin and diversity, had already 'been virtually demolished by other sources of evidence' (Harlan, 1971: 468). Moreover the designation of these *in-situ* sites of PGR as 'natural habitats' contradicts its own purifying impulse, as the differentiating potentialities of plant mutation, migration and genetic drift that mark them out entangle natural and human processes of selection (Simmonds, 1979). As an early opponent to the Undertaking from the USA was forced to acknowledge,

> A large amount of genetic material dealing specifically with crop genetic resources is not present in this form in nature. It is present in primitive forms of agriculture and it would mean preserving areas of primitive agriculture. (Bommer, Assistant Director General Agriculture, FAO, 1979: 177)

His conclusion, that it would be difficult to ask developing countries to undermine their 'modernization' efforts by protecting such 'primitive' agricultural practices, could not have been more at odds with that of delegates speaking for such countries who, like Lopez-Portillo, held just such protection as a vital defence against their ongoing erosion.

For the purposes of the Undertaking, *ex-situ* PGR refer to a network of collections storing plant germplasm removed from its 'native habitat'. These 'controlled environments' include: seedbanks; field genebanks such as arboreta, plantations and botanical gardens useful for vegetatively propagated crops and trees; and *in vitro* storage facilities that conserve plant parts, tissue or cells in a nutrient medium (FAO, 1993: 20). Among the most important such collections are those of International Agricultural Research Centres (IARCs) associated with the 'green revolution' (see Yapa, 1993), and a network of national research and storage facilities (see figure 5.4.). The distribution of these *ex-situ* genebanks is almost a mirror image of that of *in-situ* sites, with industrialized countries predominating in terms of their location, funding and management (see Wilkes, 1983).[12] Such facilities are liable to various technical and financial problems, but their biggest shortcoming is that plants thus removed from the spaces of living

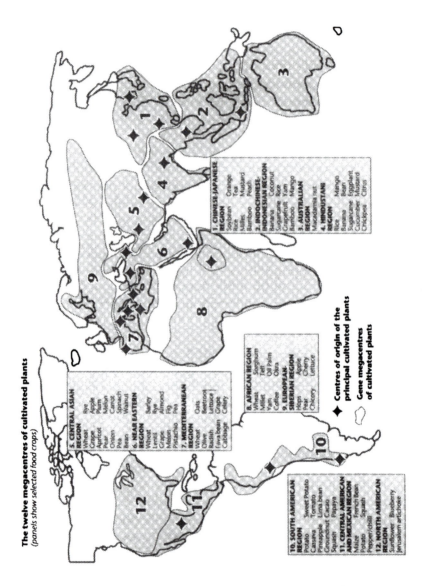

The twelve megacentres of cultivated plants
(panels show selected food crops)

2. CENTRAL ASIAN REGION
Rye
Apple
Apricot
Pear
Melon
Onion
Carrot
Pea
Spinach
Bean
Walnut

6. NEAR EASTERN REGION
Wheat
Rye
Barley
Lentil
Grape
Fig
Almond
Pistachio
Pea

7. MEDITERRANEAN REGION
Wheat
Olive
Radish
Broad bean
Cabbage
Oats
Beetroot
Lettuce
Grape
Celery

1. CHINESE-JAPANESE REGION
Soybean
Rice
Millet
Bamboo

2. INDOCHINESE-INDONESIAN REGION
Banana
Grapefruit
Tobacco

3. AUSTRALIAN REGION
Macadamia nut

4. HINDUSTANI REGION
Rice
Banana
Sugarcane
Cucumber
Chickpea

Orange
Tea
Mulberry
Peach

Coconut
Taro
Mango

Mango
Bean
Eggplant
Mustard
Citrus

5. AFRICAN REGION
Wheat
Millet
Yams
Coffee

Sorghum
Oil palm
Okra

9. EUROPEAN SIBERIAN REGION
Pear
Apple
Chicory

Apple
Cherry
Lettuce

10. SOUTH AMERICAN REGION
Potato
Cassava
Pineapple
Groundnut
Squash

Sweet Potato
Tomato
Lima bean
Cacao
Peppers

11. CENTRAL AMERICAN AND MEXICAN REGION
Maize
Potato
Pepper/chilli

French bean
Squash

12. NORTH AMERICAN REGION
Sunflower
Jerusalem artichoke

Blueberry

◆ **Centres of origin of the principal cultivated plants**

⬭ **Gene megacentres of cultivated plants**

Figure 5.3 *In situ* centres of plant genetic diversity (after Vavilov) (FAO, 1993: 8)

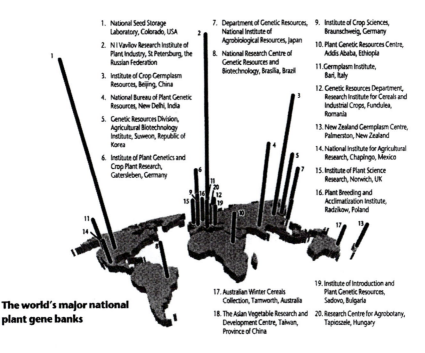

1. National Seed Storage Laboratory, Colorado, USA
2. N I Vavilov Research Institute of Plant Industry, St Petersburg, the Russian Federation
3. Institute of Crop Germplasm Resources, Beijing, China
4. National Bureau of Plant Genetic Resources, New Delhi, India
5. Genetic Resources Division, Agricultural Biotechnology Institute, Suweon, Republic of Korea
6. Institute of Plant Genetics and Crop Plant Research, Gatersleben, Germany

7. Department of Genetic Resources, National Institute of Agrobiological Resources, Japan
8. National Research Centre of Genetic Resources and Biotechnology, Brasília, Brazil

9. Institute of Crop Sciences, Braunschweig, Germany
10. Plant Genetic Resources Centre, Addis Ababa, Ethiopia
11. Germplasm Institute, Bari, Italy
12. Genetic Resources Department, Research Institute for Cereals and Industrial Crops, Fundulea, Romania
13. New Zealand Germplasm Centre, Palmerston, New Zealand
14. National Institute for Agricultural Research, Chapingo, Mexico
15. Institute of Plant Science Research, Norwich, UK
16. Plant Breeding and Acclimatization Institute, Radzikow, Poland

17. Australian Winter Cereals Collection, Tamworth, Australia
18. The Asian Vegetable Research and Development Centre, Taiwan, Province of China

19. Institute of Introduction and Plant Genetic Resources, Sadovo, Bulgaria
20. Research Centre for Agrobotany, Tapioszele, Hungary

The world's major national plant gene banks

Figure 5.4 *Ex situ* genebank collections of plant genetic diversity (FAO, 1993: 20)

association are no longer active in the process of differentiation that was the raison d'être of global governance. For delegates from industrialized countries, such sites underlined the removal of their holdings from the state of nature and should have put them beyond the collective space of the flesh. For those from developing countries, they harboured the spoils of acts of appropriation that made their incorporation into the 'global system' all the more imperative.

Making the cut between the natural and the social in the materiality of PGR is a labour of division performed neither by 'reading off' an objective demarcation between the wild and the cultivated in the flesh of plants, nor through the technical determination of scientific practices. What is remarkable, at least in hindsight, is how weakly the 'genetic' referent in PGR is aligned in the collective. In the event, this potent acronym diagrams a space of the flesh that is cellular rather than molecular in its material–semiotic habits, allying these labours to an epistemological refrain of plant germplasm and varieties that was already being outmoded by the digital calculus of DNA (see Rabinow, 1996). In this, the boundaries between PGR as a gift of nature or a social artefact return the disputations of this collective event again and again to the alignments of sovereignty and property and the legal practices that effect them.

2nd cut – between global and national jurisdictions

> Rethinking the meaning of democracy cannot be separated from a
> fundamental rethinking of the principle of state sovereignty as a key
> practice through which a specifically modern reification of spatio-
> temporal relations affirms a specifically modern answer to all ques-
> tions about who 'we' could possibly be. (Walker, 1991: 255)

If the object of the International Undertaking proved difficult to fix in the
flesh of plants, the 'global' body politic that it convened was no easier to
carve out from the practised territorialities of sovereign nation states (see
der Derian and Shapiro, 1989; Ruiz, 1991). The collective 'we' tentatively
gathering as/in this event was nothing if not ambitious – a polity of
humankind that took its cue from earlier episodes in global environmental
governance which had seen the oceans, Antarctica and outer space (re)con-
stituted through international treaties as 'the common heritage of [hu]man-
kind' (CHM) (Barrère, 1992; Buck, 1998).[13] As Wolfgang Sachs has
observed, this codification of the globe as a commons proclaims the unity
of humanity not as some enlightenment fancy but as a 'bio-physical fact',
the social corollary of 'one earth' (1993: 107). In the words of one the most
influential advocates of this global polity,

> Until recently, the planet was a large world in which human activities and
> their effects were neatly compartmentalised within nations. The tradi-
> tional forms of national sovereignty are increasingly being challenged by
> the realities of ecological and economic interdependence. (WCED Report,
> 1987: 4)

The spatial vernacular of the 'global commons' both mimics and disrupts
the enduring currency of the commons as a parable of modern political
ordering that marks its narrative compass and practical limits (Goodrich,
1991; M. Shapiro, 1991). This currency harnesses rival impulses of 'tragedy' and 'virtue' vested in earlier sitings of the commons as spaces outside
the territorializations of sovereign nations or private property, such as in
the 'backward' countryside of industrializing England (Thompson, 1991)
or the 'wilderness' of North America to European settlers (Cronon, 1983).
Where the 'tragic' impulse persists in the characterization of remnant forms
of 'pre-modern' resource management in the third world as wasteful and
unsustainable (see Hardin, 1968),[14] the global commons assume a virtuous
environmental hue as the last preserve of an original nature inherently
resistant to the contrivances of enclosure (Murphy, 1977).[15]

The International Undertaking was unambiguous in its collective
resolution to govern PGR in the name of humankind: 'This Undertaking is
based on the universally accepted principle that PGR are a heritage of
mankind and consequently should be available without restriction' (Article

1, FAO, 1983a: 50). But just as the legal import of these nice words was not readily apparent on first reading earlier in this chapter, so too was it lost in confusion and disagreement in the event. The spirit of the Resolution was best articulated by the Mexican delegate who proposed it to Conference in 1983:

> Genetic resources must effectively be considered in all respects as a heritage of mankind and in consequence, germplasm should be made freely available in all times and places. . . . Because of its importance to life, both plant and animal, this heritage must be conserved and used judiciously and carefully, which is to say, in the service of the interests and needs of all humanity. (Lopez-Portillo, FAO, 1983b: 295, author's translation from Spanish)

Opponents, like the Canadian delegate, nonetheless made plain their dissent from the assemblage of such a polity in PGR, contradicting the habit of collectivity in his opening remarks to the next meeting of Conference:

> My delegation first wishes to emphasize and put on record that, despite the impression created by the report of the 88th session of Council [CL/88/rep/1: paras. 26–31], there is no consensus in either the Council or the Conference on FAO's activities and initiatives in this area. (Fredette, FAO, 1985a: 298)

Others continued to invest these words with the same legal weight and political force as the constitutional provisions of the global commons they sought to emulate: '. . . we have been told, and in fact we know, that PGR are the common heritage of mankind. No one doubts this' (Amukoa, Kenyan Delegation, FAO, 1983b: 288). Far from being a stray mistake, this slippage between the 'heritage of mankind' and the established legal concept of CHM litters the minutes and verbatim records of Conference and Council deliberations (e.g. FAO, 1983c: para. 107).

However, as is obvious from the maps in figures 5.3 and 5.4, PGR contravene the territorial vernacular of the global commons. Unlike Antarctica or the oceans, the space of the polity cannot be constituted as an 'outside' of modern political ordering, but is thoroughly enmeshed through a dense fabric of associations with the spatial practices of national sovereignty and private property. The collective assemblage of PGR as a jurisdiction of global governance unsettles rather than reinforces these coordinates, complicating the constitutional performance of International law. In the first place, as we have seen, plant germplasm is mutable and mobile within and between efforts to 'fix' it as/in *in-situ* 'native habitats' and *ex-situ* 'collections'. Moreover, these sitings are already bound up with projects of nation building whether through the legacies of colonial science or ongoing practices of state modernization. Constituting PGR as a 'global commons' inevitably cross-cuts these territorializations, reconfiguring these categorical spaces as connected points on multiple trajectories and shifting the modality of jurisdiction to a governance of flows. Only by convening 'free access to PGR' in these terms could the geo-political interests and assets of first and third world countries be rendered common currency in/as the space of a global polity. This gathering of forces faltered at one of the boldest gestures of collective intent, the Undertaking to establish

> . . . an internationally coordinated network of national, regional and international centres, including an international network of base collections in genebanks *under the auspices and/or jurisdiction* of the FAO. (Article 7.1a, Resolution 8/83, FAO, 1983a, my emphasis)

While delegations from industrialized countries were happy to see their long-accustomed access to *in-situ* germplasm in 'developing' countries reaffirmed as a global commons, the prior claims of plant breeders' rights were repeatedly marshalled to effect the removal of *ex-situ* collections from its constitutional compass (see the 3rd cut below). As the Mexican delegate made clear, this insistence reinforced the very habits of unequal exchange that the Undertaking set out to redress:

> In an ideal world the best and easiest course would be for complete and free exchange across the international network of genebanks, but we cannot disguise or ignore the fact that until now the biggest and almost exclusive benefits of this whole process have gone to transnational companies. . . . Day by day those countries, which are the original proprietors of genetic resources, are becoming obliged to pay for our genetic material on its return from centres in industrialized countries. (Lopez-Portillo, FAO, 1985a: 343–4, author's translation)

Some measure of the persistence and intensity of this political fault-line is evident in the collective recourse in 1989 to reiterating the constitutional parameters of the Undertaking's jurisdiction by formally voting on an 'Agreed Interpretation', three Conferences after its initial adoption.[16] This Resolution (C4/89) confirmed that the 'global system'

> . . . covers the conservation and use of *ex-situ* and *in-situ* biological diversity in plant genes, genotypes and genepools at molecular, population, species and ecosystem levels. (FAO, 1989a)[17]

In the meantime, the status of *in-situ* and *ex-situ* collections in terms of their legal constitution and jurisdiction had been under prolonged investigation by FAO legal advisers (see CPGR, 1987a). They concluded that under International law *in-situ* germplasm fell within the jurisdiction of the nation state in which it was located, but that the situation for *ex-situ* collections was less clear-cut, particularly in the case of International genebanks.[18] Summarizing their advice, Lopez-Portillo spelt out the political consequences of these obscure legalities:

> ... the genetic resources collected in any country in the world and stored *ex-situ* become the legal property of the centre or the country in which the genebank is found, passing to the place of storage, irrespective of the place which originally produced it. Collections stored in national centres are under national jurisdiction and cannot guarantee free availability, notwithstanding declarations of good intent. Also collections stored in CGIAR centres, appear to be the property of each centre falling under the jurisdiction of its administrative council. (FAO, 1985a: 293, author's translation)

With industrialized countries intransigent in their opposition to the collective realignment of *ex-situ* collections under their direct or indirect jurisdiction, developing countries also began to retreat from the space of the commons and to reassert their own sovereign rights over *in-situ* PGR. One of the first to break ranks was the Ethiopian delegation, which argued that

> Genes are a valuable resource for any country. We think it is the sovereign right of that country to make use of them by any means it deems necessary and anybody who would like to acquire a genetic resource should agree on the mode of its acquisition with the proprietor [*sic*]. . . . We fail to understand why PGR should be considered any differently. Ethiopia has reservations about the International Undertaking on PGR . . . until the relevant articles are adequately amended. (Debabu, FAO, 1985a: 307)

In the absence of any positive movement on the status of *ex-situ* germplasm following the 'Agreed interpretation',[19] support swelled among delegations from developing countries for the Ethiopian position. In 1991, this realignment of forces came to a head in the form of a Resolution (C3/91), formally endorsing 'that nation's have sovereign rights over their PGR' (FAO, 1991a), and annexing this Resolution to the legal provisions of the International Undertaking.

Carving out a global jurisdiction for the governance of PGR in the name of 'humankind' was frustrated not only by political divisions between nations but by the frailties of Inter-national law in constituting such a jurisdiction amidst, rather than 'outside', the spatial practices of national sovereignty and private property. Where the protocol of the 'common heritage of mankind' (see note 12) had been instrumental in the constitution of other 'global commons', the assemblage of a commons in PGR breached the territorial vernacular of its legal provisions even as it haunted the collective labours of division. Here, the CPGR was itself party to perpetuating confusion. For example, in the preamble to Resolution 4/89 drafted by the Commission secretariat, Conference was still being invited to recognize that 'plant genetic resources are a common heritage of mankind to be preserved and made freely available for use' (FAO, 1989a: 27).[20] In practice, as the Commission acknowledged in its own working sessions, the legal status of the Undertaking to place genebank collections 'under the auspices and/or jurisdiction of the FAO' amounted to little more than a 'generic' reference to some unspecified form of 'control' (CPGR, 1987b: 4). In the event, the least line of resistance led to the adoption of the weakest of four legal options considered, which saw the FAO constituted as a 'trustee on behalf of the International Community' of germplasm collections volunteered by Governments or International bodies. This arrangement vested no beneficiary interest in, or direct managerial, administrative or financial control over, any germplasm placed 'under its auspices' (1987b: 8). At the same time, *in-situ* PGR were effectively removed from the common space of this global polity by the collective Resolution to affirm the sovereign rights of nation states. The 'heritage of mankind' proved a poor imitation of the legal principle of CHM in constituting a global jurisdiction in the governance of PGR, coming rather closer to the more modest legal provisions of the UNESCO Convention on the Protection of World Cultural and Natural Heritage 1972 (CPGR, 1994c). But this was only acknowledged in retrospect as CPGR entered negotiations to harmonize the provisions of the Undertaking with those of the Convention on Biological Diversity in 1994 and claimed, somewhat ingenuously, that there had never been any intention 'to exclude . . . either the overriding sovereign rights of a state . . . or private property rights existing under national law' (CPGR, 1994c: 6).

3rd cut – between heritage and invention

> It is a specifically western proclivity, and a late one at that, to treat innovation as a product of the intellect and the products of the intellect as separate from other aspects of the person. (Strathern, 1998: 231)

Like technology, property renders knowledge practices durable, attaching and detaching people differentially to networks of heterogeneous others – human and non-human, living and inert, whose capabilities are intricately interwoven through practice (see Anderson, 1998). As we saw in the case of 'real' property in the last chapter, property law does not simply regulate the relationship between pre-existing (natural) objects and (social) subjects but is constitutive of the division in the socio-material fabric of worldly associations that brings things and persons into being in particular ways. Human/plant relations can be 'cut' a number of ways, for example through their constitution in whole or in part (such as seeds or fruit) as objects of 'physical' property to which specific (and sometimes multiple) entitlements pertain as a consequence of their attachment to land or customary use (Correa, 1995). But the kind of property rights that have already made their mark in this collective event, the plant breeders' rights that compromised the constitution of PGR as a polity of 'humankind', are of a different order of proprietorial assemblage in the canon of western property law. 'Intangible' property describes objects 'that subsist by virtue of human mental life' in which the effects of creativity are 'transubstantiated' in the fabric of things (see Drahos, 1996).[21] Here, the space of the subject is constituted by prescribing discrete acts and singular agents (individual or corporate) of creativity from the flux of knowledge practices through the assignment of intellectual property rights (IPR), such as patents, copyright and trademarks, which entitle the holder to prevent others from producing, using, selling or importing, the designated 'knowledge-object' for a fixed period (usually 17–20 years) (Cornish, 1999). As Woodmansee's (1984) study of the fraught legal constitution of the 'author' as a title-holder in eighteenth-century copyright law illustrates so well, the subjects of property no more prefigure the conduct of law than do its objects.

For all their 'intangibility' the evidence of the knowledge-object itself is crucial to substantiating such rights and herein lies the rub for contriving proprietorial subjects through IPR in relation to *living* things. Historically, the jurisprudential distinction between 'physical' and 'intangible' property has construed living things as belonging 'by their very nature' to the domain of the physical, thereby ruling them outside the compass of IPR (Hamilton, 1993). Not until the 1980s (i.e. contemporaneous with this collective event) did legislation and case law, led by the US Supreme Court, begin to shift these ontological coordinates by making a new cut between biological and microbiological knowledge practices and objects that admitted biochemical 'in(ter)ventions' and genetic entities into the company of patentable things (Correa, 1995).[22] In these circumstances, plant breeders' rights (PBR) can be considered a 'poor relation' of IPR, assembled against the jurisprudential grain through the laborious elaboration of an international convention signed in Paris in 1961 – the Union pour la Protection des Obtentions Végétal (UPOV).[23] PBR exercise a lower threshold of 'inventiveness' than patents, tailored to the specific socio-materialities of plant propogation in the form of whole plants, seeds or other generative parts, and evidenced in/as varietal entities that have to exhibit 'stability, uniformity and distinctiveness' to substantiate entitlement (Aubertin and Vivien, 1998: 42).[24]

Industrialized nations opposed to the collective Resolution to constitute PGR as a 'heritage of humankind' were not always as forthright as the US delegation in spelling out the kinds of knowledge practices and practitioners prescribed as/by PBR:

> We recognize the concern of some Member nations regarding the effects of breeders' rights legislation on germplasm exchange. We are convinced however, that such legislation causes little or no hindrance to free germplasm exchange among plant *scientists* around the world. (Benjamin, US delegation, FAO, 1983b: 285–6, my emphasis)

If science is *the* mode of knowing convened by the assemblage of IPR as the benchmark of 'inventiveness', by the same token, the market is *the* calculus of exchange by which what is worth knowing is calibrated.

> My country is committed to free germplasm exchange that recognizes private rights. . . . Plant breeding has been a major source of new varieties to feed the world, while negating that initiative would have serious detrimental effects to farmers and consumers worldwide. In the United States alone, close to 90% of plant breeding maize is conducted by the private sector; in vegetables it is 50% or more. (Gayoso, US delegation, FAO, 1985b: 325)

For proponents of the Undertaking, it afforded an opportunity to challenge the legitimacy of this 'cut' or, as Alfons Bora puts it, to confront it with the law before the Law (1999: 145). If, as we have seen, the bodying forth of social knowledges in the fabric of plants does not begin with science, the challenge was to expose the prejudicial assumptions of western legalities that privilege its practices and artefacts and reconfigure the space of the subject in terms that did justice to other knowledge practices, practitioners and artefacts. The Pakistani delegation articulated the collective resolve in just these terms.

> Is ownership an academic question? . . . Given, in fact, a hierarchy of breeders' rights ranging from a chance discovery of a mutation to deliberate genetic engineering, which right in this hierarchy is an acceptable right and which is not? . . . Given the long and chequered history and legal philosophy of . . . patents and intellectual property concepts in general in all the developed countries over the last 100 years or more, . . . the work of the Commission should be the starting point for reaching [a better] understanding. (Musharaf, Pakistani delegation, FAO, 1985b: 329)

If carving out a global jurisdiction in PGR as a 'heritage in mankind' had exercised the territorial vernacular of modern political ordering, the labours of division in the space of the proprietorial subject were to interfere with its temporal syntax (see O'Neill, 1997). IPR combine the universalizing pretensions of science and law to affect a radical break with the past, collapsing botanical becomings into the here and now of invention such that a germplasm without history is folded into a future of monopoly entitlement. Working against the long shadow of enclosure, proponents of the Undertaking insinuated another history into the proprietorial assemblage of PGR, affording recognition to the knowledge practices of generations of farmers embodied in the flesh of plants. In perhaps the most inventive Resolution of this governing collective, Conference endorsed an unprecedented legal provision for 'farmers' rights' in PGR as part of the 'Agreed interpretation' of the Undertaking. Resolution 5/89 endorses

> the concept of Farmers' rights (Farmers' rights means rights arising from the past, present and future contributions of farmers in conserving, improving, and making available PGR, particularly those in the International Community, as trustee for present and future generations of farmers, for the purpose of ensuring full benefits to farmers, and supporting the continuation of their contributions, as well as the attainment of the overall purposes of the International Undertaking). (FAO, 1989a: 28–9)

As the Assistant Director-General of Agriculture at the FAO made clear in his introduction to the Conference debate on the Resolution, farmers' rights were conceived as a counter-balance to PBR, according 'farmers' a form of recompense for their contribution as 'donors of germplasm' comparable in principle to that accorded plant breeders as 'donors of technology' (Bonte-Friedheim, FAO, 1989b: 253). While farmers' rights were not an individual entitlement to proprietorial benefit, they did constitute a legal mechanism for financial compensation akin to a collective form of copyright protection for cultural knowledges where exclusive rights cannot be substantiated in relation to a specified object (Correa, 1995).[25] Just as importantly, they provided a formal means to stop the prior and ongoing work of generations of farmers 'disappearing into doneness' (Star, 1991b: 121), making the past permanently present as an enduring heritage and ongoing contribution and entitlement to compensation.

The moral and practical force of this principle and its historicity were accepted by some delegations from industrialized countries, like the French:

> It is clear to my delegation that the term 'farmers' must be understood in the broadest sense, that is to say, to refer in a generic way to all men and women of all countries and of both past and future generations, who cultivate the earth/land (*la terre*) for their subsistence or to get an income. (Piotet, FAO, 1989b: 264, author's translation from French)

But others found this proprietorial realignment hard to swallow, casting farmers' rights as 'antagonistic' to the prevailing (scientific) order (Australian delegation, FAO 1989b: 263). It was left to the Canadian delegate explicitly to refuse the knowledge practices recognized by these rights as constituting 'proper' knowledge at all, urging that:

> Most farmers conserve and improve plant germplasm basically for their own needs, not with the goal of conserving PGR as we understand the term today. It is therefore difficult to clearly understand what the resolution means when it refers to the concept of 'supporting the continuation of their [farmers'] contributions' in the endorsing paragraph. Surely we support the work of trusted experts . . . to best assist farmers and farming communities in the protection and conservation of PGR. (Tubino, FAO, 1989b: 269)

He even suggested alternative wording for the Resolution on 'farmers' rights' that turned the very concept on its head, replacing farmers with 'qualified groups engaged in genetic resource conservation in all regions of the world', and insisting that 'without the application of scientific knowledge, technology and financial investment to PGR, these resources are of little use to solve the pressing problems and challenges facing agriculture' (ibid.: 268).

Far from being an academic question, ownership is an intensely political and practical business of law('s)-making. The modality of invention constituted through/as IPR depends crucially on effecting a cut between a proprietorial subject and object by displacing the physicality of knowledge practices from the corporeal conduct of 'thinking' to its transaction by the artefact (see Callon, 1998), like a plant variety in the case of PBR. This is a prescription, as Vandana Shiva puts it,

> . . . for a monoculture of knowledge which displaces other ways of knowing, other objectives for knowledge creation and other modes of knowledge sharing . . . that contributes immeasurably to our intellectual and cultural impoverishment. (1993: 33)

Challenging this temporal projection of disembodied thoughts into dissociated things, the space of the subject in this governing collective complicated the historicities of socio-material attachment and exchange, admitting other knowledge practices and practitioners into the company of just entitlement. However, in its laborious balancing of powers between PBR and farmers' rights, this assemblage of PGR found itself overtaken by the pace of legal change elsewhere as courts in the USA and Europe extended patenting to the in(ter)ventions of genetic engineering (see note 22). Moreover, in making the cut between invention and heritage this governing collective reiterates the calculus of cultivation reverberating from the last chapter, drawing the proprietorial line at indigenous peoples whose livelihood practices are as precariously placed in the space of the 'developing' nation as in that of colonial 'settlement' (see Kloppenburg, 1995; Posey and Dutfield 1996).[26]

de/reterritorializing PGR

'. . . the notion of the global environment, far from marking humanity's reintegration into the world, signals the culmination of a process of separation. (Ingold, 1993: 31)

Like the deeds and fences that reconstituted human attachments to the land
by elevating 'cultivation' to a reason at law, so the impulse of enclosure is
at work here in the flesh, carving discrete acts/agents of 'invention' from the
flux of knowledge practices that body forth in the intimate fabric of plants.
Its modernizing embrace continues to draw 'an expert knife through the
carcass of custom, cutting the use-right from the user' (Thompson 1991:
134) and rendering the particularity of diverse socio-material belongings
commensurable through a universal legal standard that pegs entitlement to
the market. However, unlike those dispossessed by the colonial settlement
of the land whose absent presence in the conduct of law-making haunts the
last chapter, in the event of the Undertaking, their counterparts constitute a
majority voice in the chambers of inter-national governance. Here, the legal
conceit that hybrid plants are a product and instrument of modern science
is forced into dialogue with the subaltern historicities that it would silence,
redistributing hybridity through long and diversely practised habits of
human/plant association. In this, the collective assemblage of PGR affects a
post-colonial contra-modernity in which 'the Third World, far from being
confined to its assigned space . . . penetrate[s] the inner sanctum of the First
World in the process of being Third World-ed' (Prakash, quoted in Bhabha,
1994: 247).

These spaces and practices of division between nature and culture,
global and national, heritage and invention overlap and intermesh but are
not reducible one to the other, even as their collective alignment is a mutual
accomplishment. In tagging the documents that are the artefacts and
mediators of these labours of division we catch something of the act of
telling tales, the 'fabulations' as Deleuze would have it of a constituency in
the making (1995: 125–6). The collective event of the International Undertaking is no more an ending than a beginning in the governance of PGR.
The CPGR continues to perform the 'global system' but has spent most of
its energies since 1993 'harmonizing' its protocols and mandate with those
of the Convention on Biological Diversity, a realignment signalled by a
change of name to the Commission on Genetic Resources for Food and
Agriculture (CPGR/94/WG9/3). In these fraught negotiations, the 'cuts'
freighted by its 'soft-law' Resolutions both endure and mutate. At its eighth
session in April 1999, the urgency of convening a common global agenda
on 'agricultural biodiversity' had become a 'priority not only for universal
food and livelihood security but', as one NGO participant put it, 'a lifeline
for the FAO itself' (Mulvaney, 1999). If PGR proved an ineffective witness

to the constitution of a global commons it remains an effective boundary object in making space for human/plant associations in the more pristine 'natures' of biodiversity. If the historicities of 'heritage' have lost some of their force against the prohibitive modality of 'invention', their subaltern impulse continues to make its mark as the device of 'farmers' rights' remains in circulation, informing a proliferation of *sui generis* forms of intellectual property right in the making (Crucible Group, 1994; Posey and Dutfield, 1996).

If 'hybrids presage new projects of modernity', as Strathern suggests, they do not simply reinforce the coordinates of modernization as the space in which society begins and the 'state of nature' ends. The spaces of political community convened as/by global environmental governance refuse as much as they observe the universalizing pretensions of these modern space–time coordinates, articulating a more dis-orderly 'a-modernity' in which the pre/post/modern are all swept up (Patton, 2000). As an imagined 'space beyond modernity', the idea of the 'global commons' and the political practices being forged in its name, demonstrate an important silence in contemporary political and legal theory and their characterizations of the disintegration of the universal pretensions of Modern governance and justice. Such accounts focus on the tension between the impulses of sovereignty and exchange, where sovereignty 'tends toward . . . specifying and bounding both the spaces in which subjects achieve eligibility and those in which the collective as a whole has dominion', and exchange 'encourages flows and the relaxation of bounda-ries to produce expanded domains in which things can circulate' (Shapiro M., 1991: 448). In legal terms this returns us to the dilemmas of governing flows as/through 'a zone of rights' in which the uni-versality of the Law is central to the very idea of legal order as the negation of the 'state of nature' (Fitzpatrick, 1992: 82). But the dis-ordering of these de/territorializations inheres in the refusal of 'things' to observe the divisions that they are called upon to witness, as well as in the disputes between nations 'represented' in the event of global governance. The question of 'what democracy could possibly be in relation to "the people" in these un/re-makings of political community (Walker, 1991) is not just a question of negotiating global citizenship through the awkward vernacular of inter-national relations, but through the affects of 'things' in these cosmopolitan assemblages (Stengers, 1996).

The parable of the commons, I suggest, marks the spatial inscription of the mutually informing and contested contours of modern purifications of the inside/outside boundaries of political community through the spatial practices of sovereignty and of the social/natural boundaries of living associations through the spatial practices of property (see figure 5.5). Delimiting the narrative limits and practical possibilities of the modern political imagination, the commons parable is an unavoidable and intensely

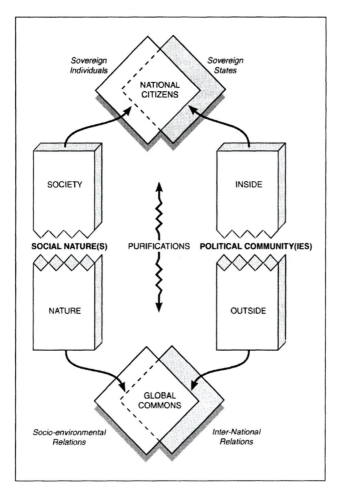

Figure 5.5 De/re-territorializing plant genetic resources

contested boundary condition of all new worlds (Rose, 1986). Dis-placing
this parable to the terrain of global environmental governance provides an
important interference in its moral topology and impetus for new political
alignments and theoretical initiatives in the spatial configuration of socio-
material associations (Helgason and Pálsson, 1997). The distant territorializations of the 'tragedy' of the local commons in the Third World and of
the 'virtue' of the pristine spaces of the global commons become impossible
intimates in the contested vernacular of biodiversity (Ostrum *et al.*, 1999).
Here, the bio-prospecting alliances of western corporations, life sciences
and states engineering new commodities out of the 'state of nature' are
compelled to engage with communities and organizations of resistance in
the South, and their Northern allies, who rightly insist that there is no 'state

of nature' only richly inhabited ecologies in which the precious metal of bio-diversity is intimately bound up with the diversity of cultural practices (Goldman, 1998).

The socio-material hybridity of PGR, and its complex geographies and histories, disturbs the territorial vernacular of global governance and harbingers new spatialities of jurisdiction and (dis)possession that promise to shift the political imaginaries of who and what constitutes 'the commons' at the beginning of the twenty-first century. The politics of 'global ecology' (Sachs, 1994) or 'earth politics' (von Weisäcker, 1994) are necessarily more plural and partial than a global vision that maps a universal subject, the 'we' of humanity, on to a powerful image of a finite terrain. What is needed, as Shiv Visvanathan suggests, is 'not a common future but [a re-imagining of] the future as a commons' (1991: 383).

Section 3

• • . ¯

The facts cut me off. The clean boxes of history, geography, science, art. What is the separateness of things when the current that flows each to each is live? It is the livingness I want. (Jeanette Winterson, 1997: 85)

Just as humans and non-humans, subjects and objects are constituted through the performance of myriad moments and modes of consumption, so the distinction between them is rendered porous in the process as 'things' become familiar co-habitants in the living fabrics of association that configure the geographies of the social. It is in these most intimate and ordinary of encounters and accommodations that the bodily contours taken to mark off discrete 'individuals' and/of certain kinds amidst the heterogenous flux of the socio-material world gain a subversive potency as liminal spaces in which self-effacing habits, tacit knowledges and embodied desires refuse the apartheid distinction between nature and society even as it continues to hold sway over the discursive conventions of expert and commonsense accounts of the world. In the essays in this section I want to turn from the spectacular spaces of wildlife and the proprietary spaces of governmentality to explore this interval between sense and sense-making in the quotidean spaces of everyday life. Here, as Arjun Appadurai puts it,

> the small habits of consumption, typically daily food habits, can perform a percussive role in organizing large-scale consumption patterns . . . made up of much more complex orders of repetition and improvisation. (1996: 68)

Appadurai's efforts are directed towards resituating the temporality of consumption in terms of the polyvalent rhythms of history, periodicity and process so as to avoid its treatment as the 'end of the road for goods and services' (1996: 66). I ally this impulse to a reconsideration of the spatiality of consumption against the confined conceits of the shopping mall and the supermarket, the archetypal predestinations of goods originated in the spaces of production. These chapters endeavour to resituate consumption in more visceral terms, incorporating the multiple sites of inhabitation connecting the bodily spaces that locate 'our' being-in-the-world to the metabolic frailties and corporeal compulsions of multifarious 'others' that share the precarious register of life and redistribute its energies through all manner of intermediaries and configurations. In this enterprise I take my cue from A.N. Whitehead's observation that 'the body is only a peculiarly intimate bit of the world' (1929: 113), to suggest that taking embodiment as the nexus of our situatedness in the world is to foreground relationality, rather than individuality, as the axiom of social life. This is a relationality, like Whitehead's 'concresances', Deleuze and Guattari's 'rhizomes' or ANT's 'networks', that necessarily extends the social beyond the human or, more properly, through which the human and other kinds are con-figured in particular and provisional ways. The inter-corporeal intimacies that body forth through everyday consumption practices are nowhere more in evidence that in that most carnal and compulsory of exchanges – nourishment (see de Certeau *et al.*, 1998).

Appadurai notes in his seminal book *The social life of things* (1986) that food is a ready candidate when it comes to appreciating the inherent tropic qualities of things. Variously mobilized as goods, products or commodities, the stuff of food constantly shifts register between the material and the metaphorical, between plant or animal and crop or livestock, between calories and bodies (ibid.: 506–7). Moreover, as Elspeth Probyn suggests, this mortal traffic 'offers a very practical figuring of an everyday ethics of living' (1999: 224) in which food complicates the geographies of intimacy, stretching and folding the time–spaces of here and now, 'us' and 'them', producing and consuming in complex and contested ways. However, I want to take issue with the mute pliancy of the objects mobilized in both of their accounts in so far as they restrict the terms on which 'things' are admitted into the social to their animation by something else (human intention), excluding the affectivity of 'things' on their own account – affects that can resist and deflect the course of human designs (see Graves-Brown, 2000: 4). My argument in the following chapters is that the lively currents in this inter-corporeal commotion amount to more than simply a 'traffic in things' set in motion by exclusively human subjects.[1] To work against the asymmetry of such 'methodological animism' is to recognize with Peter Pels

> . . . that materiality is not some quality distinguishing an object from a subject – that one should, in fact, question the slippage from the epistemological to the ontological notion of 'object' which undergirds arguments of material anthropology. . . . It implies that the 'material' is not necessarily on the receiving end of plastic power, a *tabula rasa* on which signification is conferred by humans. (1998: 100–1)

Configured in more relational terms, the corporeality of the body and of the world fold through each other in the manner suggested by Merleau-Ponty's notion of the 'reversibility of the flesh' (1968). Tracing this principle through the bodies caught up in the geographies of food simultaneously exposes the neglected pulse of prehension, sensibility and disposition in the fabric of 'human' being (and doing) and readmits what Michel Serres would call the '*sens* of objects' (1991: 13) into the company of worldly agents with whom our lives are entangled. Far from denying embodied difference, such a relational conception of social life is what makes corporeal differentiation possible by requiring us to distinguish between bodies in ways that appreciate their specificities and respect the affects and affordances that flow between them and through which each takes and holds (and changes) shape. Like Winterson, 'it is the livingness' I want to hold on to in these last chapters, by attending to the flesh of the 'things' as well as that of the (always human) subjects in the metabolic spaces of consumption, and by exploring the ethical import of their shared 'corporeal imperative' (Weiss, 1999: 128). Chapter 6 traces the polyvalent career of the soybean (*Glycine max*) as an industrial crop and a Frankenstein food through its genetic enrolment to freight the world's leading herbicide – Roundup®. One of a litany of 'food scares' to have complicated the cartography of connections within which people situate their shopping and eating habits, GMHT (genetically modified herbicide tolerant) soya disturbs the orderly transposition of socio-material value from field to plate with a more monstrous topicality. Chapter 7 takes up the ethical import of the everyday negotiation of such hybrid geographies. Public anxieties around industrial foodstuffs and growing consumer participation in alternative food networks, from Fairtrade to organics, suggest that food is a ready messenger of connectedness and considerability that is fleshing out the spaces and practices of a relational ethics, even as academic and policy analysts struggle to register or make sense of them.

Transgressing Objectivity:
the monstrous topicality of 'GM' food

Eating scrambles neat demarcations and points to the messy inter-
connection of the local and the global, the inside and the outside . . .
food compels us to think about . . . the social as a surface
composed of relations of proximity. (Elspeth Probyn, 1998: 161)

eating space

From the mundane discomforts of indigestion or the sour grimaces that
mimic the odour of foodstuffs on the turn to the collective anxieties that
aggregate around any number of toxins and diseases freighted by food,
eating marks the most immediate and commonplace enactment of Merleau-
Ponty's insistence that 'the lived body is our general medium for having a
world' (1962: 130). The metabolic impressions that the flesh of others
imparts to our own is an enduring axiom of social relations with the non-
human world and the porosity of the imagined borders which mark 'us' off
from 'it' (Douglas, 1966; Fiddes, 1991). The potency of this vector of intercorporeality seems to grow as the moments and spaces of cultivating and
eating, animal and meat, plant and fruit, become ever more convoluted.
The troubling spectres of fleshy mutability that haunt the shadowy regions
between field and plate mass with particular intensity in the event of 'food
scares'. Such events are endemic to the relentless industrialization of food
over the last half-century and are emblematic of the threadbare fabric of
trust (dis)connecting industrial food production and consumption as we
enter the twenty-first century (Griffiths and Wallace, 1998).[1]

Listeria, Salmonella, *E. coli*, dioxin, chemical, hormone and antibiotic
residues and, scariest of all perhaps, 'mad cows' are now familiar inter-
lopers in the cheap abundance and superfluous choices enjoyed by those of
us accustomed to an industrial diet. But the 'yuk factor', as Derek Burke so
felicitously labels such gut apprehensions (1998), shifts ambivalently
through this catalogue of unwelcome familiars. What begins as a catalogue

of errors by accident – 'rogue' bacteria and proteins whose presence signals a failure in the clinical production and distribution of milk and meat, chickens and eggs,[2] becomes a catalogue of errors by design – the traces of scientific and economic rationalizations of plant and animal bodies as crops and livestock that, in their multifarious incarnations as human foods, become incorporated into our own (Fitzsimmons and Goodman, 1998). Unlike Ulrich Beck's neat demarcation between 'natural hazards' and 'social risks', the benchmark that heralds his *Risk society* (1989), the social and the natural are not so readily distilled in this parade of disturbing hybridities. Rather, such food scares freight all manner of metabolic histories and geographies – the subconscious patina of eating taboos; the medical registration of allergies and illness; the rampant mutability of cellular life; and discursive ruptures in the credibility of 'experts'. It is hard to imagine a less propitious context for the clandestine advent of genetically modified (GM) foods – the more monstrous in their topicality for being undetectable by texture, smell or appearance in the field and positively unrecognizable on the plate, anonymous ingredients in the welter of processed foods that passes our lips.

Incredible as it seems with hindsight, the corporate and state agencies most implicated in the fabric of industrial agri-food networks behaved as if they were unaware of, or indifferent to, the potency of this dissonance between popular apprehensions of the hyphenated spaces between growing and eating and their own polished assertions that the hyphen is incidental, conveying nothing significant – 'trust us'.[3] This disjunction in spatial imaginaries is nicely illustrated by the juxtaposition of two humorous depictions of the turbid interval between field and plate. The first is an advertisement from the trade press in the late 1980s by one of the then leading US agri-biotechnology companies, Arco Seed (see figure 6.1a). Under the byline 'Our taste is a product of culture' is an image of a silver platter and candelabras, the accoutrements of fine dining. The glow of the candles reflects seductively off the platter on which an assortment of petri dishes is served, their coded numbers visible beneath a decorative flourish of parsley. Here the space–time coordinates of growing and eating are neatly conjugated. The company's technical claims to accelerate the corporeal improvement of crop plants are allied to a re-location of food's tastiness from the cultural sites of the restaurant/kitchen to the 'tissue culture' of the laboratory. Yield and taste are identical scientific achievements. The second depiction is a cartoon from the French newspaper *France Soir* in the late 1990s (see figure 6.1b). Its caption is 'Le secret alimentaire', a pun in which the secrecy of industrial foodstuffs is what keeps the potboiler of publicity simmering. The cartoon collides the cosy gender relations of a mealtime scene, a woman bringing food to a table at which a man is seated expectantly, with a tersely disconcerting exchange of

Figure 6.1a Arco Seed Company advertisement from the late 1980s: 'Our taste is a product of culture' (Busch *et al.*, 1991: 227)

Figure 6.1b Cartoon from *France Soir* from the late 1990s: 'Le secret alimentaire' (Cartoonist: Trez; *France Soir* 26/2/99)

words. His clichéd enquiry 'What's for supper?' is met by the food-bearer's guileless reply, 'If only I knew?'

More than any previous food scare, the abbreviated opacity of GM has provided a vehicle for articulating this diffuse but mounting sense among food consumers that they no longer know what they are eating or, just as significantly, trust authorities claiming to know better (Durant, 1998; ESRC, 1999). Implicit in this gathering unease is Probyn's observation at the start of this chapter that eating complicates the clean cut spatialities of local/global, inside/outside and public/private and forces us to engage rather different geographical imaginations. But, where her analysis of the McDonaldization of family and civic identities suggests a liberating disruption to the stifling gender relations that are performed through household food preparation and consumption routines, not everything is rosy in the post/modern hypermarket. In the case of food scares in general, and GM foods in particular, collective glimpses of the unfamiliar folds of laboratories and corporate headquarters, law courts and government offices that complicate the straight line from field to shelf are rather more politically fraught. A case in point is taking place in Britain. Located between the food cultures of the USA and continental Europe and an epicentre of the tragic blunder of BSE–vCJD, Britain has become an unlikely hotbed of civil and consumer defiance of the scientific and political authorities that look after 'the public interest' in food matters and have been so conspicuously compromised (Hinchliffe, 2001).

However, just as governments and corporations have been slow to acknowledge the misfit between their own logistical cartographies of food and the more intimate geographies inhabited by consumers, so too have the research accounts of social scientists. The topological 'compulsion' that Probyn attributes to eating has been widely resisted in agri-food studies which have tended to fracture along an economic/cultural fault-line and, through their conversations and alignments with political economy and cultural studies respectively, to reiterate the compartmentalization of production and consumption (see Goodman, 1999). The staple concepts of agri-food studies, such as *commodity chains* (Friedland *et al.*, 1981); *filières agro-alimentaires* (Allaire and Boyar, 1989) and *systems of provision* (Fine *et al.*, 1996), share a tendency to configure the geographies of food as a unilateral translation of socio-material value from field to plate, in which food is little more than the terminus of the crop.[4] If 'consumption' has been something of an afterthought in these studies, cultural approaches have been just as circumscribed in their attentions. While they have succeeded in animating food consumption as a socially complex and consequential process, their focus on shopping, cooking and eating identities and the bodily register of these cultural practices (Lupton, 1996; Bell and Valentine, 1997) rarely strays much beyond the supermarket aisles, restaurant tables and take-away menus where food, it appears, is replicated at will.[5] Everything that matters in these bi-partite accounts of the geographies of food seems to boil down to profitability or subjectivity. The *matter* of agri-food becomes an absent presence, like the hyphen that holds the moments of producing and consuming in place forgetting, as does Probyn, that the traffic between them is a traffic in and through 'things' (see Jackson, 1999).

As Probyn suggests, food, of all things, complicates these well-worn distances between production and consumption in ways which should 'render visible the lines of force that produce the pleats and folds of our social lives' (1996: 11). But the vegetal currency of Deleuzian philosophy to which she appeals in this manoeuvre does more than alert us to the spatial import of its rhizomatic diagramming between here and there, now and then. It also complicates the fabric of 'our social lives' in ways that Probyn does not interrogate, by extending the register of 'bodies' that count beyond the human and admitting *living* things and their traces (not restricted to the visible) into this vital topology. Evocations of the rhizome in social theory rehearse its botanical identity as an 'underground stem' that produces branches under the soil surface from which shoots emerge above ground, extending a plant's reach in space and time. But, while such evocations readily grasp its metaphorical purchase – the kinds of spatialities *the* rhizome 'stands for', they rarely hold on to its material potency – what rhizomes 'do' and the living spaces they inhabit (Buchanan, 1997). This is a potency more commonly appreciated in the practical knowledges of gardeners and growers through their dealings with clumping or invasive plants like ferns and lilies, or potatoes (Rost *et al.*, 1998: 101 and 399).

Thus, even as accounts of food consumption have become populated by 'body-subjects', these fleshy concerns remain resolutely human in scope.[6] Yet it is the rhizome's cellular plasticity that informs its figurative use in Deleuzian bio-philosophy – the energetic exchanges and becomings of plants and insects, plant cells and soil microbes, water and light that are aligned in variable and heterogeneous ways against the genealogical unity of the tree, which centres everything on itself and is rooted to the spot (see Ansell-Pearson, 1999: 160–1). For Deleuze and Guattari, the rhizome freights spatial practices and imaginations in which 'the variability, the polyvocality of directions, is an essential feature' (1988: 382).[7] In place of the straight lines and orderly sequences of food chains, *filières* and systems which project originary points of production through frictionless trajectories to terminal points of consumption, the geographies mapped here are more turbulent and more attentive to the multiplicity of possible paths 'in-between' where things pick up speed and take on consistencies and directions of their own.

In this chapter I ally this rhizome figure to forays into the lively worlds of *Leguminosae* – a family of some 18,000 species of plants identifiable taxonomically by the shared trait of a fruiting pod (Polhill and Raven, 1981), but more popularly known by their most colourful feature, a pea-like flower. *Leguminosae* constitute nearly one-twelfth of all known flowering plants and are second only to grasses (cereals) in terms of their economic significance as food crops (ILDIS, 2000). More particularly,

I want to weave two moments in the rhizomatic geographies of a particular legume – the soybean (*Glycine max*). In its manifold guises as seed, plant, bean, oil, flour and emulsifier (lecithin) (the list goes on),[8] soya takes here the analytical part of Serres' 'excluded middle/third' (*le tiers exclu*) (1995); a blank figure which announces the presence of absent 'things' in the fabric of social orderings and compels us to attend to the significance of their being left out of the narratives of social analysis (Hetherington and Lee, 2000).[9] As such, it transacts the crop/food fault-line not simply by its motility and indeterminacy but in its tendency to change the conditions of possibility, the valency of connections along the way (Munro, 1997).

The two moments in the rhizomatic geographies of GM soya that I trace below are those of its becoming an industrial crop and a Frankenstein food. As industrial crop, the soybean is the artefact of energetic associations between plants and people in which ecological adaptation, seed selection and plant breeding have all left their mark on its agronomic properties. In its GM incarnation the soybean has become one of a number of transgenic crops fabricated under the trademark Roundup Ready™ that have been genetically enrolled to tolerate a broad spectrum (i.e. indiscriminate) glyphosate herbicide Roundup®. The crops and herbicides that bear the Roundup logo are produced and marketed by Monsanto, one of the largest agri-chemical corporations in the world.[10] As Frankenstein food, soya is among the most ubiquitous and discreet components of industrial diets with two of its derivatives – soya flour and soya oil – finding their way into a host of processed foodstuffs from margarine, confectionary and soft-drinks to take-away and oven-ready meals. Here, soya galvanizes hectic currents of anxiety about the surreptitious presence of transgenic materials in the things we eat into improbable lines of force that are even now realigning everyday eating habits and the organizational practices of food retailers, manufacturers and government agencies. These moments of becoming 'genetically modified' pervert the nutritional configuration of the soybean's social qualities as a protein-rich foodstuff and re-align its socio-material valency in peculiar and contradictory ways.[11]

becoming industrial soybean

> . . . to enrol animals, plants, proteins in the emerging collective, one must first endow them with the social characteristics necessary for their integration. (Bruno Latour, 1994b: 60)

For the connoisseur, the collection of 7,359 soybeans held at the University of Illinois at Urbana-Champaign is a worthy exhibition of the object of their arcane enthusiasm (Hapgood, 1987). For the rest of us it is an obscure reminder that even as we take 'the' soybean for granted as a certain kind it is as manifestly heterogeneous and fraught a socio-material assemblage as any other in the migrant fabric of the USA. However intimately it has come to be fixed as a scientific object, under the taxon *Glycine max*[12] or the microscopic scrutiny of the cellular anatomy of its proteins, lipids and root nodules, the soybean retains the energetic and variable propensities of plantlife. But such natural histories occlude much older ones which have witnessed the soybean's enrolment in social relations as a cultivated crop over many centuries. Like wheat, maize, sorghum, potato and other staple food plants, the complex relations between human, animal and plant communities that are abridged in the notion of domestication have all left their mark on the soybean's transposition from its place as a regional cultivar in north-east China for more than 3,000 years to that of an industrial crop in north America since the early twentieth century (Hancock, 1992).[13] Three glimpses of the social enrolment and shifting potencies of the soybean signal the laborious and volatile directionalities of this journey.

'brings happiness'

The versatile growing and eating properties of soybeans made them highly prized among peasant farmers, acquiring familiar names like 'brings happiness' and 'yellow jewel', and saw them circulating as a dietary staple throughout East Asia by 1,000 AD in a variety of guises from sprouts and beans, to processed derivatives like bean curd and fermented pastes and sauces (Kiple and Ornelas, 2000). These are properties known today in terms of their nutritional value as a major source of plant protein and their capacity to fix nitrogen from the soil, and so thrive in relatively poor conditions. Like many plants united under the taxon *Leguminosae*,[14] the soybean's nitrogen-fixing capability is associated with its distinctive root nodules, which are themselves an expression of thoroughly symbiotic relations between the plant's root cells and soil bacteria. Under the lens of twentieth-century science these metabolic intimacies are writ large (see figure 6.2) in the enzymatic conversion of soil nitrogen by the bacterium *Rhizobium* sp. and its journey through a tiny infection thread in the plant's root hairs into the cortex of the root to form a swollen mass of cells (the nodule), drawing energy all the while from the carbohyrdates provided by the host plant (see Hirsch and LaRue, 1997).

The dense fabric of socio-material relations between plants and farmers, soils and bacteria accumulated in Asia has engendered thousands of genetically and phenotypically variable 'land races' of soybean with different environmental and disease tolerances and susceptibilties associated with their particular ecological nexus (Frankel and Bennett, 1970).[15] These range from indigestable flat-lying 'wild' varieties to those habituated by selection to stand upright and bear larger more useful seeds and beans

Figure 6.2 The soybean's nitrogen fixing root nodule (adapted from Rost *et al.*, 1998: 82, figure 5.20)

(Fowler and Mooney, 1990). The sub-species *Glycine max*, the accepted taxon of the modern cultivated soybean, is itself a product of these ancient associations, whose ancestry has been traced back to the 'wild' sub-species *Glycine soja* (Bao *et al.*, 1993). Thus, while the soybean's hybridity has been intensified and realigned by recent transgenic engineerings of its cellular DNA, it does not originate in them anymore than it is the outcome of purely 'natural' hybridization (see Rieseberg and Ellstrand, 1993). But, if it was the soybean's properties of nourishing both soils and people that freighted its early enlistment into the social networks and practical knowledges of Chinese peasant farmers, its enrolment as an industrial crop articulates a very different kind of social ordering.

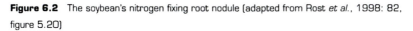

'magical hybrids'

Accounts of the soybean's arrival in North America celebrate a seaman-merchant Samuel Bowen as the founding figure who, on his return from speculative travels in China in 1765, brought the first seeds to a plantation in Savannah, Georgia (Hymowitz and Harlan, 1983).[16] But little of consequence seems to have followed his intervention in terms of the soybean's career as an industrial crop. In Jack Kloppenberg's (1988) compelling history of the slow and fraught accretion of diverse acts through which the biotechnological transformation of agricultural plants has proceeded, the soybean's reassemblage as an industrial commodity emerges as anything but straightforward. Neither scientific 'proofs' of its extraordinary nutritional value published in France in the late 1880s (Hapgood, 1987), nor new cooking techniques imported from China in the 1930s to improve its palatability (Lappé and Bailey, 1999), had much effect on establishing a soybean market in the west.

Not until 1900 did it become enmeshed in the hectic scientific, governmental and commercial networks configuring the institutional fabric of US agri-food industrialization (see Kenney, 1986; Busch *et al.*, 1991). Over the next 30 years the soybean was targeted by a Department of Agriculture 'introduction programme' that enlisted scientific, diplomatic and naval energies in garnering over 4,000 varieties from across East Asia, monitoring and selecting them in university experimental stations nationwide, and introducing the most successful ones into the US farming repertoire (Evans, 1998). By 1924 a soybean crop of some 2.5 million acres was recorded in the USA, mainly in the mid-west, with a market value as animal fodder of US$24 million (Kiple and Ornelas, 2000).[17] But the soybean was proving a recalcitrant industrial subject not only in terms of its commercial performance as a foodstuff, but also in terms of the obstacles it presented to the new plant breeding techniques that promised to realize the potential of the seed itself as an industrial commodity.

This promise took shape in the socio-technical project of hybridization which was to become the talisman of agronomic science in the middle decades of the twentieth century. As Henry Munger, a plant breeder writing for the American Seed Association in 1952 put it, 'the word "hybrid" has magic in it at the present time' (quoted in Kloppenberg, 1988: 124). Its spell was first made flesh in the guise of F_1 (first generation) hybrids which codified the scientific and legal practices of Mendelian genetics and Plant Breeders' Rights in the name of crop improvement (see Fowler and Mooney, 1990). The prototype for these early efforts in plant hybridization was corn (*Zea mays*). The distinctive configuration of its male (tassle) and female (silk) flower parts, which predisposes it to cross-fertilize with other plants (outbreeding), made corn uniquely amenable to manual hybridization methods (Tudge, 1993: 178). Intervening in its fertilization, and hence in the traits inherited by the next generation of plants, involved a laborious but fairly straightforward process of pollination by hand (Goodman *et al.*, 1987: 39).[18] Standing about two feet tall and bearing seeds in fuzzy pods clustered near the stalk, the soybean (like the majority of crop plants) is inbreeding with its flower parts anatomically arranged for self-fertilization, making such hybridization techniques inpracticable (see figure 6.3).

As Mendel's experiments with garden peas had suggested decades before, the 'improved' traits exhibited in F_1 plant hybrids do not pass to subsequent generations (Tudge, 1993: 176). The ancient practice of saving

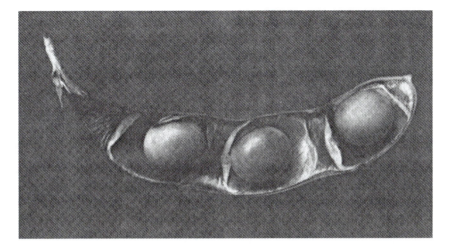

Figure 6.3 *Glycine max*: the soybean in the pod (Photo: Chris Johns; *National Geographic*, 172/1: 72, 1987)

seed from the harvest to sow next year's crop became redundant as hybridization required new F_1 seed to be purchased afresh each year. Notwithstanding the uncertain benefits of hybrid corn varieties (Berlan and Lewontin, 1986: 788), government monies poured into campuses and institutes world-wide to extend the technical purchase of reproductive hybridization to a wider range of crop plants (Busch *et al.*, 1991: 61–2). But the unprecedented potential to turn seeds into commodities in their own right required new legal instruments to secure the commercial viability of these techniques. In the case of the soybean a coincidence of technical and legal developments in the 1970s more effectively harnessed its energies to this industrial social ordering. First, the axiom of hybridization shifted from sexual intervention in the pollination process to manipulation of the process of cellular division (mitosis), so-called tissue culture methods, which increased the speed and ambit of hybrid development (Tudge, 1993).[19] Secondly, the Plant Varieties Protection Act that came into force in the USA in 1970, and was matched by parallel legislation elsewhere, provided the legal instruments to inscribe the seeds that embodied these socio-technical processes as discrete and attributable social artefacts (UPOV, 1972). The acreage and value of soybeans rose sharply in the USA through the 1970s and 1980s as research and development shifted to private companies like Jacob Hartz Seed, the market leader in hybrid soybean varieties. Moreover, increasing demand for vegetable oils as a staple ingredient in processed foods enhanced the commodity value of soya, and other palm-oil alternatives (Kloppenberg, 1988: 143). However, as with many hybrid cultivars, the genetic impoverishment of the industrial soybean increased its susceptiblity to all manner of pests and diseases, while its mechanization as a monocultural crop magnified the economic significance of 'weeds' as a contaminating presence in the field. Alongside the sterile seed, hybridization had bred a remedial appetite for technical fixes.

Roundup Ready™

By 1998 some 60 million acres of soybeans were being grown in the USA with a harvest value of almost $12 billion, dwarfing production in Asia and accounting for half of the total global crop.[20] More remarkable still, as a measure of its intensifying industrialization, nearly a third of this crop consisted of Monsanto's Roundup Ready™ soybean, a genetically modified herbicide tolerant (GMHT) variety that had only been licensed for commercial planting in the USA in 1996 (Food and Drink Federation, 1998). This startling entrance of GM soybeans had nothing to do with improving the nutritional properties or commercial value of the bean. Rather it signals the increasingly monopolistic impetus of corporate efforts to enrol the seed into the service of other product lines in the agro-industrial stable. Thus over 70 per cent of the current acreage of GM crops world-wide is accounted for by herbicide-resistant varieties (ISAAA, 1998). Existing hybridization techniques were quite capable of breeding pest- and disease-resistance traits garnered from the diversity of the Asian soybean land races into the branded hybrids of the US industrial crop (Tudge, 1993: 191). But the genetic modification of the soybean (and other crop plants) presented a quicker and much more commercially attractive vehicle for businesses like Monsanto, whose investment in hybrid seeds was fuelled by their established interests in agri-chemicals (Kloppenberg, 1988). In this latest industrial twist, the soybean was to be realigned to freight the world's leading herbicide, a glyphosate-based weed-killer developed and patented by Monsanto in the 1970s under the Roundup® label that accounts for 17 per cent of their $9 billion annual sales today (Anderson L., 1999) (see figure 6.4).[21]

If corn had been the prototype for early industrial hybrid plant breeding techniques which treated plant varieties as plastic groups subject to manipulation by cross-pollination and tissue culture, the tobacco plant (*Nicotinina tobacum*) was to take its place in the biotechnology era in which even these vestiges of difference were to be unstitched at the molecular level through genetic engineering. Here, as Monsanto puts it on its public information webpage,

> The DNA from different organisms is essentially the same – simply a set
> of instructions that direct cells to make the proteins that are the basis of

Figure 6.4 Roundup Ready™ crop notice in an Illinois field (*The Guardian*, 20/2/99)

> life. Whether the DNA is from a micro-organism, a plant, an animal or a
> human, it is made of the same materials. (Monsanto, 2000)

This matter of fact rendition of the highly contested dogma of molecular
genetics reduces the generative properties of all biological organisms to a
function of the genetic 'programme' encoded in their DNA (see Webster
and Goodwin, 1996; Lewontin, 1998). But it is a rendition that also
conveys the social import of this digital calculus in terms of the potential
for the 'code' to be 'mastered', unconstrained by the biological syntax of
morphogenic integrity.

Biotechnological plant breeding 'recombines' the DNA of the target
plant by altering its genetic sequence, or in the case of transgenic plants, by
adding one or more genes from a donor organism (Watson *et al.*, 1992).
This recombinant (rDNA) process involves three key steps. The isolation of
the coding sequence for the gene(s) associated with the desired trait
(identification); the replication and transfer of this gene to plant cells
(T-DNA vector construction); and the regeneration and developmental
regulation of the gene in the target plant (propogation and expression
control) using conventional tissue culture techniques (Schmidt, 1995).
Early experiments to confer glyphosate tolerance to plants in the 1980s
involved the identification of the *aroA* gene from a glyphosate-resistent
mutagenized strain of the *Salmonella typhimurium* bacteria, its integration

into tobacco plant cells via an *Agrobacterium rhizogenes* T-DNA vector and subsequent expression as a modest increase in the regenerated plants' glyphosate tolerance (Comai *et al.*, 1985a, 1985b).[22]

In their corporate publicity, Monsanto present this rDNA process as a straightforward extension of traditional plant breeding methods that simply 'allows for the transfer of a greater variety of genetic information in a more precise, controlled manner' (Monsanto, 2000). As well as being a matter of dispute within the life science community, this disarming concatenation belies the arduous business of experimental trial and error that perturbs any veneer of 'controlled precision'. Practitioners are all too aware that each of these steps is fraught with all manner of technical, behavioural and legal gliches that yield uncertain results and measure 'success' over a 9–12-year time-horizon from experimentation to commercial crop production (Mazur, 1995). The influential work of Comai *et al.* (1985a), for example, reported that, alongside increased glyphosate tolerance, their GM tobacco plants exhibited 30 per cent impairment of growth over untreated control plants. The commercial impetus to which scientific careers and innovations in this field are harnessed, together with the obstinate specificities of particular target plants and vectoral organisms, tend to be written out of textbook accounts of rDNA procedures (e.g. Watson *et al.*, 1992: chapter 24). Their only trace in such catalogues of rDNA achievement is the rapidity with which standardized accounts are outmoded as, for example, with the circumvention of the vectoral step in biolistic methods (see Walden and Wingender, 1995).

In their Roundup Ready™ incarnation soybeans are hybrid agents of corporate science in which the entanglement of technical and business practices is incorporated in the seed. In technical terms, they manifest an *Agrobacterium*-mediated gene transfer of mutant forms of EPSPS enzymes (see note 21) from glyphosate-resistant soil bacteria *Pseudomonas* spp. and *Klebsiella pneumoniae* which contain amino-acid substitutions that counteract the suppression of the plant's own EPSPS by glyphosate and so increase the soybean's tolerance level compared to that of unmodified varieties (Hinchee *et al.*, 1988). Only after several years of field trials was a commercially viable balance struck between gains in glyphosate tolerance and losses in yield (Delannay *et al.*, 1991). But these technical achievements are themselves performed through proprietary alignments of germplasm, instrumentation and expertise. In this case, Monsanto's acquisition of Hartz Seed Inc. and collaboration with (and eventual takeover of) Agracetus Inc. in the 1980s secured a vital alliance between the Hartz™ catalogue of hybrid soybean seeds and germplasm (Kloppenberg, 1988) and the 'Accell' gene gun technology of Agracetus Inc. (McCabe *et al.*, 1988), to achieve pole position in the GMHT seed market.[23] It is a socio-material ordering held in place by monopoly patents whose grip is reinscribed by the signature that seals every purchase agreement each time a farmer buys Roundup Ready™ seed.

> *The Grower agrees not to supply any of this seed to anyone for planting and agrees not to save any crop produced from this seed for replanting or supply saved seed to anyone for replanting. The grower agrees not to use this seed or provide it to anyone for crop breeding, research, or seed production. If a herbicide containing the same active ingredient as Roundup® Ultra herbicide (or one with a similar mode of action) is used over the top of Roundup Ready™ soybeans, the Grower agrees to use only the Roundup® branded herbicide.* (Purchase log report form, Hartz™ Seed Company; quoted in Lappé and Bailey, 1999: 53)[24]

Amidst the profusion of proteins and patents, viruses and devices that bind the soybean so intimately to this industrial assemblage, its performance still cannot be relied upon as, like other GM organisms, its lively potencies find expression in deviant and unintended directions.[25] The Roundup®/Roundup Ready™ package is promoted as 'an intelligent environment-friendly solution' because glyphosate 'breaks down quickly in the soil' and, under test conditions, GMHT soybeans require lower treatment levels which 'experts anticipate' means that herbicide use 'can be reduced by a third' (Monsanto, 1996). Unsettling such carefully worded claims is evidence that metabolizing high levels of glyphosate induces physiological and biochemical changes in soya plants and microbial soil organisms in ways which inhibit their beneficial interaction and raise phyto-oestrogen levels in the plant that may pose a potential health risk to humans and other mammals (Nottingham, 1998; Raganarsdottir, 2000).[26] Moreover, the first two-year study by the USDA (United States Department of Agriculture) of its on-farm use in conjunction with GMHT soya, cotton and maize found that there was no measurable reduction in glyphosate application in more than half the study's regional crop combinations (7/12) and no significant difference in yield over non-GMHT crops in two-thirds of regional crop combinations (12/18) (*New Scientist*, 10/7/99). For all its precision engineering, then, the GM incarnation of the soybean no more stays put in the germinal fabric of the seed or the field boundaries of the crop than did its forebear 'brings happiness', but is metabolized and redistributed through all manner of inter-corporeal relations in growing and eating practices.

Becoming frankenstein food

> These ways of everyday practice, these procedures and ruses of consumers constitute the network of an antidiscipline. (de Certeau, 1988: 14)

The arrival of the first shipment of 200,000 tons of US soybeans incorporating Roundup Ready™ GMHT beans in Europe in October 1996 met with a hostile reception. German Greenpeace demonstrators sporting white 'biohazard' jumpsuits and rabbit masks outside the Unilever headquarters in Hamburg, one of Europe's largest soybean importers, made headlines with placards protesting 'we don't want to be laboratory rabbits' and 'no to GM soybeans in our food'. So began what Time Magazine (1996) called the 'battle of the bean genes', which has since become an energetic theatre of environmental and consumer resistence to GM foods across Europe, not least in Britain. Here, it did not take the 'fourth estate' long to displace the rabbit mask with that most English touchstone of anxiety – the monstrous visage of Mary Shelley's Frankenstein (Turney, 1998; Bingham, 2001), distilling disparate currents of apprehension into the tabloid slogan 'Frankenstein foods' (*Daily Mail*, 4/5/96). It is a visage that has come to haunt GM foods in ways quite as 'magical' in its hold on the public imagination as hybrid seeds had been as a totem of scientific progress in the 1950s (see figure 6.5).

The proliferation of small acts of refusal performed through diverse registers and spaces of consumption from supermarket tills and school meals to the diffuse vectors of popular sense-making, attest to the practical potencies of doing and knowing that de Certeau evokes in his notion of everyday life (see Conley, 1997: 111). These quotidian spaces disturb the

Figure 6.5   Frankenstein as the popular face of GM protests in the UK (*The Observer,* 21/2/99)

socio-material valency of commodities, and the strategic cartographies that they freight, by otherwise inhabiting agri-food networks and realigning the properties of foodstuffs as they transact the interval between production and consumption. Such countervailing currents in the topologies of growing and eating are not peculiar to food scares. But these events condense the volatility of industrial orderings through the metabolic intimacies that their fleshy traffic forges in its convoluted journeys from field to fork.

Over 40 per cent of the US soybean crop is exported, primarily to Europe (and Japan) where the soybean is little grown. As a consequence, it has fallen to other crops in the Roundup Ready™ stable, like oil-seed rape and sugar beet, and other brands of GMHT maize, to focus concerns here about the genetic contamination of related 'wild' species and cultivated varieties and the impacts of broad spectrum herbicide use on biodiversity (English Nature, 1998).[27] But some 60 per cent of Europe's processed foods contain soya derivatives and much of the public anxiety about GM foods has been channelled through the soybean precisely because its surreptitious presence is emblematic of the lack of advisory labelling and precautionary testing that have attended GM's arrival on the menu.[28] In other words, somewhere between field and plate the precious rDNA proteins that proponents promise will 'feed the world' cease to matter enough to entitle those destined to eat them to a means of discerning their presence or, thereby, exercising a practical judgement of any kind. In this section I will argue that this vanishing trick has fuelled the vacuum in trust which Frankenstein foods have come to occupy. Having outlined the regulatory protocols of 'substantial equivalence' which contrive a dis-junction in the GMHT soybean's transition from crop to food, I turn to consider the 'ruses and procedures of consumers' to recuperate the socio-material connections between growing and eating despite, or even because of, the wedge that such protocols would drive between them.

substantial equivalence

In their efforts to transact the interval between field and plate Monsanto gave a new twist to the promotional strategies of biotechnology companies like Arco Seed Inc. (see figure 6.1a), that had erased the hyphen holding agri-food networks in place by rendering crop and food 'improvements' coincidental scientific achievements. To counter mounting public opposition to its GMHT soybeans, the corporation launched a £3 million publicity campaign in the British press in the autumn of 1998, including seven full-page text advertisements. The first, under the byline 'More biotechnology plants mean less industrial ones', promoted Monsanto's 'environmental' claims for GMHT technologies as a departure from the damaging chemical regime of 'industrial' crops. One of its sequels, entitled

'This strawberry tastes just like a strawberry', suggested that such momentous changes to the crop in the field left no trace in the food on your plate, which tasted just the same. The hyphen that had united the spaces of production and consumption here marked a radical break, assuming the proportions of the ocean traversed by the vessels shipping soybeans from the USA to Europe.[29]

From the first portentous cargo's arrival in Antwerp, and in marked contrast to the scrupulous segregation of Roundup Ready™ soybean seed, US growers' and shippers' organizations joined Monsanto in insisting that the separation of GMHT beans from conventional ones was unnecessary because 'they are no different' and, in any event, was 'technically' impracticable (American Soybean Association, 1996). These claims were attuned to the global food safety protocols assembled by the *Codex Alimentarius* Commission under the auspices of the United Nations and encoded in its documentary proceedings (FAO/WHO, 1996). These protocols calibrate the 'risk' of GM foods by assessing their difference from 'conventional' comparators against a normative benchmark of 'substantial equivalence'. In both theory and practice the determination of 'equivalence' depends on the comparator species/variety selected, for which *Codex* provides few criteria, and the parameters and specification of testing required to substantiate equivalence, which *Codex* restricts to genetic composition (see Ho and Steinbrecher, 1998; Nature, 1999). Furthermore, claims to 'substantial equivalence' are based on evidence supplied by the applicant in advance of independent testing, thereby circumscribing regulatory scrutiny.[30] Thus, Monsanto's application to the UK Advisory Committee on Novel Foods and Processes (ACNFP) in 1994 for a licence to sell Roundup Ready™ soybeans anticipated the exemption of their product from regulatory scrutiny on the grounds that 'following the principles for the application of substantial equivalence, there should be no further safety or nutritional concerns of any significance' (quoted in Anderson L., 1999: 16).

At the time of Monsanto's application, the risk assessment of GM foodstuffs (as opposed to crops) in the UK was divided between two main Government advisory committees reporting to the Department of Health and the Ministry of Agriculture, Fisheries and Food. As well as the scrutiny of the ACNFP, responsible for the safety of novel foodstuffs, it fell within the purview of the Food Advisory Committee (FAC) responsible for food labelling (Levidow and Carr, 1996). As a House of Commons Environmental Audit Committee report later complained, these bodies (and their agronomic counterparts) were dominated by scientific and industry 'experts' party to the technical and commercial thrust of genetic engineering. Moreover, it observed that their remits were so narrowly defined and insulated that many of the 'big' questions fuelling public debate, such as the impact of GM crops on biodiversity and what actual benefits their derivatives offered consumers, fell between the respective regulatory competences of these committees and so remained effectively unaddressed (EAC, 1999).

The ACNFP accepted Monsanto's 'substantial equivalence' case and issued a licence for Roundup Ready™ soybeans in 1995 on the understanding that their transgenic rDNA components were sufficiently degraded by the processing of beans into soya meal and soya oil to effectively remove them from the human food chain.[31] Shortly afterwards the FAC, which had begun drawing up guidelines for the labelling of GM foods in 1990 (MAFF, 1990) and had undertaken a consultation exercise on the issue in 1993, advised the Government that products containing Roundup Ready™ soya required no consumer labelling since these soya derivatives had been determined to be 'substantially equivalent' to non-GM ingredients. This apparently seamless silencing of the advent of GMHT soya might have provoked little more than the well-honed protests of consumer and environmentalist organizations were it not for Monsanto's loudly declared determination to handle, transport and process its transgenic soybeans commingled with conventional ones. Consumers were not only to be deprived of the means of making a judgement but, by enlisting the entire soybean market to the Roundup Ready™ network, it became evident that they were to be denied any alternative. The 'objective' considerations and 'technical' criteria that are the currency of regulatory authority, not least in the minds of advisory committee members themselves, now appeared uncomfortably aligned with a market strategy that undermined consumer choice (Levidow, 1999).[32] With misplaced parliamentary assurances on the health risks associated with BSE still in the public spotlight, and other European and Australasian countries adopting tougher stances on the labelling of GM products, the British Government found 'substantial equivalence' a poor bulwark against the GMHT soybean's disorderly potency as Frankenstein food.

nourishing disorder

The invisible ubiquity of soya, together with the corporate disregard for consumer choice heralded by its Roundup Ready™ incarnation, galvanized disparate and, sometimes, contradictory currents of public anxiety about GM foods into a concert of civic and consumer resistance performed through, and fuelled by, intensive mass media attention (Hargreaves, 2000). Resistance manifested itself in all manner of carnivalesque protests, internet traffic and organized direct action. Newspapers whose political allegiances, journalistic styles and readerships had little else in common, jostled to champion the 'mood of the nation' with sustained campaigns to expose any number of elliptical relations between politicians, policy advisers or scientists and bio-technology corporations. Equally unlikely allies emerged among public figures who lent their voices to the anti-GM chorus, from the Prince of Wales's efforts to muster the 'moral majority' against 'tampering with the natural order' (e.g. *Daily Mail*, 1/6/99) to the cosmopolitan appeal of a coalition of celebrity chefs decrying the threatened impoverishment of European food cultures (e.g. *Independent on Sunday*, 7/2/99).

While such conspicuous activities captured the headlines, the more sobering signals of hardening public resistance to GM foods accumulated from innumerable people exercising nothing more than the prosaic power of choosing what to put in their shopping trolleys or to feed their children, and what not. First registered by supermarket check-outs and customer relations, these everyday acts of consumer resistance congealed into obstinate 'facts' in the commercial and political landscape through the pervasive metric of public opinion polls, questionnaire surveys and focus groups (Porter, 1995).[33] Faced with such inconvenient figures as '76 per cent of European consumers demand labelling of GM foods (Eurobarometer, 1996 reported in *Nature*, 1997. 387:845–7)', scientists, politicians and corporate spokespeople sought vainly to contain their import by dismissing them as 'irrational' and ascribing them to public 'ignorance' and/or media misrepresentation of GM technologies (Boulter, 1997; ESRC, 1999). However, those whose corporate fortunes are more closely attuned to 'customer behaviour' in the food market, particularly the major retail chains which had worked hard to ally themselves with 'the consumer interest' in Britain (Marsden *et al.*, 1999) and were accustomed to dictating terms to their suppliers (Competition Commission, 2000), began to take them very seriously indeed.

Far from assuaging consumer anxieties about GM foods, the sleek alignment of industrial and regulatory practices to the contrivance of a radical dis-junction between the production and consumption of Monsanto's Roundup Ready™ soya only fuelled scepticism and disbelief. Each practised reassurance – 'there is no evidence that . . .' – touched collective nerves exposed by the BSE–vCJD débâcle, nourishing a vernacular that gave voice to people's gut apprehensions of the metabolic intimacies connecting field and plate. The interval between GM crop and food that Monsanto's advertising campaign would convince us is an empty blank, instead became full of half-truths, shady practices, unknown quantities and potentialities incarnate in the soybean (Consumers' Association, 1999). Among the many disorderly currents of anxiety that it has been transducing into the countervailing forces of Frankenstein food, two stand out. The first is a chronic distrust of the established arbiters of risk that thrives on echoed remembrances of mendacious Government assurances and unreliable scientific expertise, and their compromising associations with corporate business (Rothman *et al.*, 1996; Grove-White *et al.*, 1997). The second is a wary recognition of the fleshy mutability of living things which exceeds the designs of science or commerce and warrants a more precautionary approach than that exhibited in hindsight by other food scares (Mayer *et al.*, 1996).

Some of the most instructive distillations of this visceral vernacular are to be found in the subversive humour of cartoons which animate the monstrous topicality of Frankenstein foods by disturbing the corporeal archetypes that affirm the borders of, and between, 'humans' and 'things'. In the first of the examples below (figure 6.6a) distrust of GM foods is made flesh not only in the startling size of the maize kernels on display but also the two-headed grocer trying to persuade a woman shopper to overcome her evident doubts about buying them. In the second example (figure 6.6b) the lively potencies of 'things' themselves are embodied in a transgenic vegetable that contradicts the dark-suited authority of a man assuring his audience that 'there's absolutely nothing wrong' with the GM potato he is holding, by silently identifying itself as a carrot.

This vernacular insistence on the connectivities between crop and food, gained practical momentum as food retailers harnessed it to their corporate efforts to reassert themselves in the commodity markets by subjecting GMHT soybeans to the disciplines of 'product traceability' (Valceschini, 1998).[34] Product traceability transacts the interval between production and consumption by tracing the bio-graphies of foodstuffs from field to shelf through networks of protocols, devices and personnel and rendering these journeys legible on the things they have become at point of sale by means of product labelling (Murdoch *et al.*, 2000). It was a

Figure 6.6a 'The two-headed grocer' (Cartoonist: Colin Wheeler; *The Guardian*, 1999)

Figure 6.6b 'The GM vegetable answers back' (Cartoonist: Austin; *The Guardian*, 12/2/99)

marketing strategy that had been sanctioned and promoted by the Government in efforts to 're-build' public trust in British beef in the wake of the BSE–vCJD débâcle. But in the policy vacuum of 'substantial equivalence', it was the corporate food retailers that took the initiative in the case of GMHT soya.

Sainsburys, one of the UK's leading food retailers, had been lobbying the US soybean industry and the British Government 'behind the scenes' as early as 1995 to reverse the policy of non-segregation and non-labelling (Austin, 1999). In the absence of progress, the 'big five' companies accounting for some 80 per cent of the British food retail market followed the example of one of their smaller competitors, Iceland, in March 1998 by banning GM foods and ingredients from their own brand products (www.icelandfreeshop.com). As the company's technical director explained in an interview for a special supplement of *The Independent* newspaper on GM foods:

> We're not opposed to biotechnology but it was clearly wrong that these products were coming into the market and consumers were not being given a choice. As a retailer we could probably have made lots of money out of the technology but when we asked customers, 77% didn't want us to start selling GM foods. . . . Later we met the biotech industry and they basically said that I was a backward European who didn't like change. They tried to suggest there were no issues with environmental impact and food safety. They were so arrogant that I came back determined to do something about it. (Bill Wadsworth, 12/10/99: 4)

The retailers' strategy had a domino effect on the more reluctant food processing industry as corporations like Nestlé and Unilever, fearing that their branded products would lose market share, announced that they too would not use GM ingredients.[35] By May 1999 a European consortium of leading food retailers and processors had formed to secure supplies of non-GM ingredients and derivatives (ENDS, 1999). As they gathered pace, these countervailing commercial currents boosted the market for non-GM soya, primarily from Canada and Brazil, inflating the price and volume of sales of soybeans guaranteed *not* to be Roundup Ready™. In the process, this realignment of beans, contracts and devices that could tell and keep GM and non-GM soya apart undermined the rubric of 'equivalence' and dispelled the 'impracticablity' of their distinction.[36]

The British Government found its initial political embarrassment at being seen to side with the US soybean industry rather than UK consumers compounded by the initiatives of food retailers at home, and the readiness of key politicians in the European Union to champion the prerogative of 'consumer choice' (Fischler, 1997). In 1997, the European Commission agreed new definitions for novel food regulations (EC No. 258/97) requiring all products which 'may contain or may consist of genetically modified organisms' to be labelled and sold separately. Because GMHT soya products had been approved prior to this regulation coming into force, an amendment was passed (No. 1139/98) specifically to include derivatives of GMHT soya (and maize) within its scope (European Community, 1998). Isolated, the Government finally shifted its public stance on GM foods in 1999. The Prime Minister, who had been one of their loudest advocates, announced through a newspaper article bearing his name that he now accepted that

> There's no doubt that there is potential for harm, both in terms of human safety and in the diversity of our environment from GM food and crops . . . [and that these are] cause for legitimate public concern. (*Independent on Sunday*, 27/2/00: 28)

Shortly afterwards, in March, amendments to domestic legislation (the Food Labelling (Amendment) Regulations) to bring UK food labelling provisions in line with those of Europe came into force. In April, the Government announced sweeping changes to the remits, membership criteria and organization of the regulatory infrastructure for GM crops and foods. Two new 'strategic advisory commissions' were set up to oversee and 'join-up' the activities of the existing advisory bodies – the Agriculture, Environment and Biotechnology Commission in the case of crops and the Food Standards Agency for foodstuffs. Existing advisory committee members with links to the biotechnology industry were ceremoniously replaced with more 'independent' members representing a wider range of environmental, consumer and ethical concerns and expertise (OST, 1999).[37] As an anti-disciplinary configuration of the soybean, Frankenstein food has been proving a no less forceful or polyvalent current in the rhizomatic geographies of this extraordinary legume than that of its industrialization.

geographies in/of the flesh

> [N]ot only are humans as material as the material they mold, but humans themselves are molded, through their sensuousness, by the 'dead' matter with which they are surrounded. (Peter Pels, 1998: 101)

The unlikely assortment of sites and sources from petri dishes to shopping trolleys, technical papers and tabloid articles that I have spun together here is as promiscuous in its affiliations as the monstrous topicality of the soybean and as eclectic in its assemblage as the odd genetic bits and pieces stitched together in the idiom of rDNA. My purpose in focusing on soya itself has been to try to grasp agri-food networks not at one end or the other, where everything has always already been settled, but in-between, where so much happens but very little is taken into account. Instead of the blank figure that haunts the spatial imaginaries of commodity chains and consumer cultures in agri-food studies, the soybean emerges as a lively presence that agglomerates very diverse acts and complicates the distribution of powers and knowledges in the precarious business of growing and eating. As industrial crop it incorporates and resists efforts to discipline its germinal energies to freight a patented brand of herbicide and, thereby, perpetuate the monopoly impulse which would otherwise have expired with the patent at the end of the twentieth century. As Frankenstein food the GMHT soybean's immanent transgression of the confines of the seed and the crop transduces consumer apprehensions of the metabolic intimacies between growing and eating into anti-disciplinary alliances with environmental organizations, farming unions and retail corporations. Journeying in multifarious guises, the soybean fleshes out the interval between these distant but simultaneous moments, tracing rents and folds, currents and frictions in the topological performance of producing and consuming, global and local, 'humans' and 'things'.

In both cases, as industrial crop and Frankenstein food, the molar integrity of soya is redistributed analytically as these moments of becoming are recognized as relational achievements performed through particular socio-technological orderings, bodily metabolisms and patterns of association, just as it decomposes morphologically in its own time. However, the soybean's assemblage as a socio-material hybrid does not begin (or end) with its GMHT incarnation. In pushing its hybridity back through the 'magical' first generation hybrids of industrial plant breeding and more ancient domestic familiars like 'brings happiness' we neither arrive at some timeless germ of original 'nature', to which those who condemn GM as 'unnatural' would have us return, nor at its antipode whence to admit the social into the fabric of the soybean is to render its GM configuration indistinguishable from any other in the generic impress of cultivation (see Ingold, 2000a: 77–88). As Roundup Ready™ the soybean performs the socio-material relations of growing and nourishing very differently from those of its Asian kin. Its biological diversity and dietary profusion are depleted to a catalogue of varieties that realigns these heterogeneous potencies as the attributes of scientific inventiveness and reconfigures its hybridity as a mark of ownership (see Callon and Law, 1995). The more minutely the soybean's nourishing properties as legume and protein are 'mapped into knowledge', as Stengers (1997: 118) puts it, the less these seem to matter in this re-assemblage, as the plant's intricate symbiosis with soil bacteria is subsumed by the metabolics of glyphosate and its exceptional nutritional qualities are reduced to animal fodder, the detritus of oil extraction. But for all their universalizing ambition, the strategic and analytic cartographies made flesh in the GMHT soybean are partial and provisional. Partial, in that they co-exist with other modes of ordering, not least in contemporary Asian networks where the soybean's agricultural and dietary embeddedness endures and still accounts for a third of the protein intake of the human population today (Bao *et al.*, 1993). Provisional, in that resistence in Europe is already complicating the geographies of soya's planting, handling and shipping by strengthening the non-GMHT commodity market and organic food networks where the protocols of certification preclude GM ingredients (Soil Association, 1999).

In tracing the polyvalent currents and affects of the soybean's becoming industrial crop and Frankenstein food, one no more crosses a borderline marking off fact from fiction than one arrives at the purified domains of nature and society (see Taussig, 1993). Amplifying the painful lessons of previous food scares, most notably 'mad cow disease', the GMHT soybean has complicated this distinction and the distribution of 'ir/rationalities' in important ways. The imbroglio of anti-disciplinary knowledges and practices that it articulates as Frankenstein food is more promiscuous, but no less complex, than that incarnate in Roundup Ready™. Where the rhizomatic geographies of earlier food scares had reiterated farming and regulatory practices and spaces as prominent landmarks in the visceral mappings of consumer apprehension, the GMHT soybean has brought less familiar agencies implicated in the pleats and folds of industrial agri-food networks into far-reaching proximity. In its wake, the routine business of biotechnology companies, university laboratories, experimental field sites and government advisory panels through which GM crops and foods are being assembled has been subject to unprecedented public scrutiny. In place of the 'irrational' anxieties of a public 'ignorant of the facts', we find the reasonable doubts of publics that have learnt to be sceptical of those who authorize them and a gathering appreciation of the disputatious nature of the 'facts' themselves (see, for example, Krebs and Kacelnick, 1997; O'Riordan, 1999).[38]

Between the hyperbole of 'feeding the world', the obscurity of 'substantial equivalence' and the secrecy of 'commercial confidentiality', public concerns over GM foods struggle to come to terms with scientific disagreements over the interpretation of 'the facts' and with the partiality of their production in terms of the kinds of questions and answers they evince when scientific energies are so thoroughly enmeshed with those of commerce and governance (see Myerson, 2000). These are lessons that have been hard learned and cut deep in the British (and European) body politic. Those whose authority they challenge will find no refuge from them in the customary repertoire of fortifying the self-evidentiality of 'the facts', clothing their own judgements in the mantle of 'objectivity', or decrying the wayward influence of 'cultural, subjective and irrational factors' on the irreverent masses who call them into question (Boulter, 1997: 247). As well as unsettling the provenance of 'the facts' and any prior claim to their allegiance, the Frankenstein effect disturbs the lumpen contours of 'the public', urgently reminding us that 'there are . . . no masses, there are only ways of seeing people as masses' (Williams, 1989: 11).

The topological compulsion that Probyn attributes to eating resonates with everyday practical knowledges more readily than the strategic and analytic calculi that plot straight lines from field to plate. Food scares, like foods themselves, are apprehended viscerally as well as cognitively by people remote from the business of growing and food production but alive to the transgressive liminality of foodstuffs in transducing energies, sensations and diseases between bodies. It is not that such corporeal sensibilities anchor perception and action in some primordial way, but rather, in the manner of Merleau-Ponty's notion of corporeal intentionality, that cognition itself works through them – never fully escaping its bodily inherence (1968: 149–55). The tacit skills of smell, touch and taste body forth in the habits and discernments of food consumers amidst the proliferation of instructions like branding, nutritional labelling, or best-before dates. In the case of GM foods, and GMHT soya derivatives specifically, rendering these skills redundant by claiming that consumers wouldn't be able to 'taste the difference' was as ill-judged a response to public apprehensions as it was disempowering. Rather than disqualify them from the compass of knowledge practices that bear on what we eat and the topology of connections made in the eating, they are a vital check on the monopoly of reason, nourishing other possibilities, rationalities and judgements to which scientific methods may be allied without being vested with the final word.

It is this matrix of knowledge practices that must be re-aligned if the overstretched fabric of trust transacting the distant intimacies of growing and eating is to be rewoven. As the interval between these moments becomes both more complicated and more legible through events like food scares, the burden of trust requires new and more reliable intermediaries than those aligned in the Roundup Ready™ soybean. The re-assemblage of trust in industrial foodstuffs through practices like product traceability, itself indebted to the protocols of alternative food networks such as organics and Fairtrade (Whatmore and Thorne, 1997), complicates the directionality and valency of forces between plate and field in important ways. Not least, such practices enlarge the repertoire of 'facts' and the company of those aligned in their production such that the place of materiality in social life is redistributed through the metabolic geographies freighted by food in terms of the bodily situatedness of being-in-the-world, the volatile assemblage of living kinds and the vital traffic of energies between them. Of all 'things', GM foods attest to the relational configuration of the social and the material, subjects and objects, in which the 'dead matter' of things refashioned through rDNA technologies transgress their objectivity, harbouring other possibilities than the designs of those who fabricate them, and reminding us that we too are candidates for objectification. In this, they force us to attend to ways in which the non-human makes its presence felt in social life and to find ways of registering these messengers in our accounts of the world which help us learn 'to laugh at and make others laugh at reductionist strategies' (Stengers, 1997: 90), like the cartoon, perhaps, in which the vegetable answers back.

7

Geographies of/for a More Than Human World:

towards a relational ethics

> Through exclusively social contracts, we have abandoned the bond
> that connects us to the world. . . . What language do the things of
> the world speak that we might come to an understanding of them
> contractually? . . . In fact, the Earth speaks to us in terms of forces,
> bonds and interactions . . . each of the partners in symbiosis thus
> owes . . . life to the other, on pain of death. (Michel Serres, 1995:
> 39)

the place of ethics

The modernist ideals of universal democracy and justice realized through
legislative regimes centred on individual rights have been the subject of
sustained feminist and environmentalist critiques, reinvigorating political
and philosophical interest in the question of ethics. Feminist writing has
focused on deconstructing the discourse of rights, highlighting the gendered
(and racialized) character of the autonomous self configured as a rights-
bearing citizen of a sovereign state (see, for example, Cornell, 1985). By
contrast, environmentalist work has centred on extending the political and
discursive economy of rights to non-human beings, challenging established
concepts of personhood and subject-status (see, for example, Callicott,
1979). These efforts share parallel concerns to establish relational, as
opposed to individual, understandings of ethical agency and to recognize
the significance of embodied, as against abstract capacities, in shaping
ethical competence and considerability. Such concerns highlight the power
of the geographical imaginaries of traditional ethical discourses and the
difficulties of disrupting the entrenched cartographies of the nation,
the neighbourhood and the individual in fashioning new possibilities for
conviviality.

Earlier chapters in this book have already attended to particular issues arising from these concerns, for example the ethical status of creatures like leopards, crocodiles and elephants as they were mobilized in the Roman arena and by conservation science today (chapters 2 and 3) and peoples rendered 'primitive' and incomplete 'persons' by the political orderings of European colonialism and their legacies for the constitution of universal human rights and institutions of global citizenship at the end of the twentieth century (chapters 4 and 5). This final chapter picks up these themes to explore what I take to be creative tensions between feminist and environmentalist efforts to empower those eclipsed in conventional ethical discourse and the company of ethical subjects which it (re)iterates. I trace some of the ways in which the conceptual and institutional parameters of the civic constitution of the self as citizen, central to feminist concerns, intersect with the humanist constitution of the subject as person at the heart of environmentalist concerns. In both cases, although for different reasons, I argue that dilemmas encountered by these attempts to construct alternative ethical orderings are intimately bound up with their observance, even in critique, of categorical distinctions between the cultural and the natural, the social and the material, the human and the non-human parodied in Michel Serres' evocation of the world estranged by the terms of Rousseau's 'social contract' (see Assad, 1999: 158). In this, they remain complicit in the humanist presumptions that characterize the discursive economy of ethics struggling, like so many other forlorn sorties across these impossible borders, to smuggle some semblance of the messy heterogeneity of being-in-the-world back into their accounts of it. I go on to suggest a number of consequences for instituting a relational understanding of ethical considerability and affect, using the mundane example of eating to shift from a discursive to a performative register which emphasizes the importance of corporeality and hybridity as modes of conduct for (re)assembling the spatial praxis of ethics in more than human terms.

homo ethicus

Ethical discourse has conventionally been framed in terms of an opposition between natural law and social contract traditions, centred on competing accounts of the primacy of 'human nature' as against civic order as the foundational claim to ethical competence and considerability (Poole, 1991). Commonly misunderstood as some kind of unchanging normative code inscribed in the heavens or the genes, natural law theories evoke the capacity for reason as the definitive basis of a distinctively human ethical standing. Early modern reinterpretations of a classical legacy, notably in the work of Locke, shifted accounts of this distinctively human capacity from the evocation of a 'common good' – the cluster of obligations generated by the patterns of interdependence in human social life – to that of an 'individual good' – the result of voluntary transactions between independent social agents.[1] The most important implication of this shift was to elevate the 'moral significance of the separateness of persons' (Buckle, 1991: 168). The emergence of the individual as the axiom of modern society is inscribed in legal, political and religious institutions and discourses. Since Kant, this founding figure of the autonomous self has been most strongly associated with the social contract tradition of ethics (Kymlicka, 1991). However, it is worth emphasizing that it is less the significance accorded to this figure that marks out the social contract tradition than the resolution it reaches for the social regulation of such individuals. Natural law resolutions rely on some underlying uniformities (of reasonableness) that can sustain the idea of universal (natural) human goods and values. Social contract resolutions rest on particular social institutions of contract (market) and rights (law) as the basis for establishing universal (impartial) 'laws of reason' as the pre-condition of ethical agency.

Contemporary elaborations of these debates can be seen in the philosophical and legal dilemmas of squaring claims to human rights with those to civil rights. The one represents a species claim to the possession of reasoning faculties as the basis for the universal ethical considerability of individuals by virtue of their constitution as human beings; the other, a political appeal to these reasoning faculties as the basis for the ethical considerability of individuals by virtue of their constitution as civic persons (McHugh, 1992). Historical changes in the legal encoding of such claims underline the unstable and disputed social meaning of both 'human' and 'person' as ethical subjects, for example in the treatment of women and non-European peoples, instabilities which persist in the treatment of children and those deemed mentally 'unfit'. Despite these dilemmas, the figure of the Cartesian individual as a pre-social vessel of abstract reason and will, memorably captured in Latour's image of the 'mind-in-a-vat' (1999a: 4), continues to dominate the terms of ethical debate.[2] Ethical agency is reduced to the impartial and universal enactment of instrumental reason, or 'enlightened self-interest', institutionalized as a contractual polity of equivalent self-present individuals divested of difference, context or circumstance.[3] Such accounts of ethical agency rely upon spatially and temporally stable conceptions of individual and collective social life – the sovereignty of self and state – etched in the cartographies of the citizen and the nation. Ironically, as Ross Poole suggests, in so far as 'the modern world revolves around the autonomous self and the sovereign state it has also destroyed the conditions of their reproduction, reducing community to an infinitely expanded network of market interactions' (1991: 141).

The commoditization of socio-material relations has disrupted this configuration of political and ethical community on two fronts. First, by eroding the territorialized authority of the nation state to govern increasingly global networks and mobilities of people and goods. Ethical communities bounded by national borders have become unsustainable because 'the nation state is no longer able to resolve the contradictions between citizenship and humanity through claims to absolute authority' (Walker, 1991: 256). Secondly, the expansion of market relations has also undermined the personalized jurisdiction of the individual citizen over a coherent domain of the self (Giddens, 1991). As Haraway has observed,

> the proper state for a western person is to have ownership of the self, to have and hold a core identity, as if it were a possession. . . . Not to have property of the self is not to be a subject and so not to have agency. (1991a: 135)

However, this private domain of the rights-bearing citizen has long been exposed as a masculine conceit. This has translated in different space–times into the dispossession of women, poor people and black people of political and ethical agency in their own right through, for example, their 'contractual' guises as wives, servants and slaves (Pateman, 1989).[4] Moreover, this extended domain of the patriarchal self underpinning liberal citizenship, the domain of the family and household, has itself become increasingly friable (Gobetti, 1992). In short, the reliability of this political and ethical constitution has become increasingly unconvincing as its spatial encoding in the separate realms of public and private (civic and domestic) competence has been progressively undermined by the disciplines of the market and the state.

Recent work in the field of political philosophy is dominated by two divergent responses to the limitations of the liberal conception of political and ethical community sketched above.[5] The first echoes a longstanding communitarian tradition which predicates the capacity to participate as ethically and politically competent subjects on the material satisfaction of 'basic human needs'. As Porter puts it:

> A concern for persons in their own right is not possible where the primacy of rights relies on an atomist conception of the self-sufficient individual. This notion maintains that human capacities need no particular social context in which to develop and hence is not attached to other normative principles concerning what is good for humans or conducive to their development. (1991: 127)

The more sophisticated communitarian accounts elaborated by writers like Sandel and Macintyre, appeal to an inter-subjective conception of the self as the basis of ethical agency. This conception seeks to qualify the absolute distinction between self and other associated with the figure of the sovereign individual 'by allowing that, in certain moral circumstances, the relevant description of the self may embrace more than a single empirically-individuated human being' (Sandel, 1982: 79–80). This set of responses has become politically influential with so-called 'new communitarianism' colouring the rhetoric of conventional political opponents of free market liberalism, like Blair's 'New Labour' Party in Britain and its disavowel of the infamous Thatcherite dictum that 'there is no such thing as society'. In its concern with the material pre-conditions of human life, this perspective re-engages with natural law arguments that ethical considerability precedes formal rights, requiring answers to the question 'rights for what?' At the same time it re-admits, in a limited way, non-human figures into the landscape of ethical community as necessary material 'resources' to service human needs. The environmental implications of this perspective were rehearsed in former US Vice-President Al Gore's populist manifesto *Earth in balance*, in which he argues that

> we have tilted so far toward individual rights and so far away from any
> sense of obligation that it is now difficult to muster an adequate defence
> of any rights vested in the community at large or in the nation – much
> less rights properly vested in all humankind. (1992: 278)

A second response to contemporary dilemmas in the conception and practice of ethical community is that associated with a broader critique of the foundational coordinates of modern society identified with post-modernism (Squires, 1993). Such critiques engage in a radical deconstruction of the twin sovereignties of self and state. Here 'the individual' is transformed into a site of multiple and fluid social identities, a repertoire which can be creatively mobilized to 'liberate' oneself from a singular or given subject-position. Among the more sustained expositions of this post-modernist interpretation of political and ethical agency is Laclau and Mouffe's project of 'radical democracy' characterized as 'a polyphony of voices, each of which constructs its own irreducible discursive identity' (1985: 191). Far from challenging the primacy of the individual as the ethical subject, this approach seems to me to reinscribe the Cartesian subject, merely replacing abstract reason with abstract desire or will. It shifts the ground of ethical and political community from conventional practices of contract between universally equivalent agents, to communicative practices of dialogue between radically different (but still exclusively human) agents.[6] While the bio-graphing individual evoked in this post-modern vision liberates the possibilities of ethical community from the involuntary associations of birth or propinquity, it does so by dislocating the promise of dialogic engagement from any vestige of the fleshy business of living.

The tensions between contractarian and natural law theories of ethical competence and considerability mark ongoing dilemmas over the relationship between social rationality and human mortality. The reified figure of the autonomous individual represents a cipher of abstract reason/will which inscribes the binaries of mind/body, self/other, subject/object on to the very possibility of ethical agency in modern society. Recent critiques from communitarian and post-modernist positions envision new possibilities but without interrogating or departing from the humanist presuppositions of the ethical discourses with which they are engaged. Communitarian approaches re-assert the situatedness of the individual and point to the inter-subjective constitution of ethical agency. However, they tend to do so by invoking normative configurations of community, like the family, the neighbourhood and the nation, without examining the power relations they enact. Moreover, this 'situatedness' is defined in terms of relations between people. Where they are addressed at all, relations with the rest of the world are treated as passive contextual extensions of human well-being. By contrast, a post-modern insistence on the radical instability of the individual tends to evoke highly disembodied, as well as disembedded, social agents (O'Neill, 1985; Pile and Thrift, 1995). In a world populated by such amorphous figures, constituted from cognitive and linguistic possibilities unshackled by the corporeal baggage of living, 'the question of what human be-ing is' (Porter, 1991: 16) becomes vacuous.

Emerging at the confluence of these various encounters with the intellectual and practical dilemmas of ethical agency is a re-cognition of formal justice as a derivative of more substantive moral propositions and ethical claims. Increasingly, this has been accompanied by a creative re-engagement with ideas of human nature not in terms of some ineluctable essence of humanity, but in terms of the predicament of finitude, the inherent decay and mortality of all living beings. As Cornell has put it, only 'by coming to terms with finitude can we gain the humility necessary to overcome the hubris of individualism' (1985: 338). Bauman's exploration of the ethical implications of mortality (1992), Giddens notion of 'life politics' (1991) and Beck's account of 'risk society' (1989) all exemplify an unprecedented interest in corporeality for understanding ethical considerability and conduct. Exploring issues such as the legal determination of the status and rights of the foetus and the medical certification of the condition of death, these writers suggest that the more reflexively we 'make ourselves' as persons the more significant bodily awareness becomes to the performance of the social, heightening the sense of shared existential vulnerability and finitude as a modality of political association and ethical recognition.

Such efforts echo everyday sensibilities and struggles to register connectivities between, for example, environmental degradation, animal welfare, and human health and well-being, often in the face of the pinched rationalities of public policy-making and the authority of scientific expertise (Hampson and Reppy, 1996). As we saw in the previous chapter, 'food scares' have become one of the most potent touchstones of such apprehensions about the uncertainties of human being, rendering them in the flesh through events like the trans-species carnage of so-called 'mad cow disease' and the clandestine arrival of genetically modified foods. These themes have been taken up most persistently and powerfully by those seeking to challenge the masculinist/humanist fantasy of an abstracted world of equivalent moral agents, most notably in feminist and environmentalist movements and critiques. These challenges centre respectively on concerns with the embodiment of difference and rationality and with the ethical significance of 'non-human' life forms and processes. In the following sections, I draw out what I see as key issues, tensions and shortcomings in these alternative accounts for the elaboration of a more relational understanding of ethical considerability and conduct.

feminist ethics: the embodiment of care?

> When identities become pure, exclusive, innocent, the potential for diverse and democratic collectivities is threatened. We are all others of invention, otherness should not be reified but used as one fertile resource of feminist solidarity. (Caraway, 1991: 1)

The celebration of difference in post-modern theories has been both highly influential and disputatious in feminist political thinking. A number of writers (see, for example, Ebert, 1991; Hennessy, 1993) distinguish between two different clusters of feminist engagements with this issue. The first, identified as 'ludic post-modernism', seeks to disrupt naturalized conceptions of identity as a model for political practice and locates the politics of difference in the discursive play of imagined possibilities in a theatre of volatile subject-positions (exemplified by the work of Mouffe, Young and Flax). The second, identified as 'resistance post-modernism', locates the politics of difference not as the effect of rhetorical strategies but of social struggles which ground the meanings contested in such strategies in the materialities of everyday living (exemplified by the work of Benhabib, Cornell and Grosz). While the distinction between these feminist accounts of a politics of difference is overdrawn and even somewhat caricatured, it points up an important area of dispute about how difference and its political (and ethical) import is constituted and understood. Echoing tensions in Nietzsche's writing, Diprose outlines the parameters of this dispute in terms of whether we are more likely to 'find our-selves' by looking inwards in an autonomous project of creative self-fabrication, or by looking outwards to our effects and relations with others as they configure our place in the world (1994: 87).

The first of these approaches employs individualist theories of difference, or what Kruks has called 'an epistemology of provenance' (1995: 4), to fashion self-exploration as a political process in itself while relying on an unspoken normative claim to the ethical equivalence of all 'subject-positions' in this privatized polity. Collective claims to political agency and ethical considerability tend to be looked upon askance, as intrinsically 'anti-difference' (for example, see Young, 1989). This leaves feminism as a political project precariously positioned by what Anderson (1992) calls the 'double gesture' of simultaneously asserting the theoretical universalism of decentred subjectivity while resorting to the practical lie of strategic essentialism to secure a space for women to identify common cause at all. Ironically, as she points out:

> the idea of subject-positions . . . precludes the possibility of an inter-subjective perspective that would define the human subject not as purely autonomous and self-present, nor as a mere place on intersecting grids, but as constituted through its ongoing relations to others . . . (Anderson, 1992: 78)

It is the second of the feminist encounters with post-modern theories which is the more suggestive to me as a means of negotiating the impasse of individualism in reconstructions of ethical community. It centres on a notion of difference-in-relation, as inter-subjectively constituted in the context of practical or lived configurations of self and community. In place of abstract or cognitive criteria, these always/already existing configurations of self and community are 'defined by contingent and particular social attachments whose moral force consists partly in the fact that living by them is inseparable from understanding ourselves as the particular persons we are' (Friedman, 1989: 278). This approach shares post-structuralist suspicions of the liberal ambition of value homogeneity, but remains committed to a practice of participatory communalism enacted through particular economic, political, scientific and civic orderings which condition individual capacities and arenas for action. As a feminist enterprise, it represents an attempt to understand the discursive construction of 'woman' across multiple modalities of difference by adopting a problematic which can trace the connections between discursive practices and the exploitative social orderings of meaning, being and struggle which permit and encode them (hooks, 1990).

The ethical dimensions of this approach are best captured in Benhabib's distinction between generalized and concrete others (1987). The generalized other stands for a universal principle of equal considerability in the right to be heard, to participate, to make a difference. The concrete other stands for more immediately realized ethical principles – of care, friendship, intimacy, solidarity and empathy which involve practical and often asymmetrical enactments of responsibility. However, Benhabib's elaboration of this inter-subjective conception of ethical agency reproduces the Habermasian error of according a priveleged status to the abstract qualities of rationality and language in the theory of 'communicative action'. In an

important step towards a more situated and practical approach to understanding ethical inter-subjectivity, Kruks argues that we should 'begin from
the situation of an embodied and practically engaged self; . . . from what
human beings do in the world . . . so as to rediscover the totality of [her/his]
practical bonds with others' (1995: 11–12). While this conception of a
materially-situated self has wider significance for the reconfiguration of
ethical community, to which I shall return later in my consideration of
environmental ethics, here I want to pursue two persistent themes in
feminist ethical thinking with which it resonates most suggestively. These
are the interconnected issues of corporeality (by which I mean both the
embodiment and mortality of living being) and the praxis of care.

Feminist concerns with the material situatedness of social identity and
of the particularity of sexed being have impelled a sustained consideration
of the politics of corporeality. These concerns have centred on the specificities of women's experiences as (potential) child-bearers, the objectification
of women's bodies, and the signification of 'woman' as nature incarnate
(Plumwood, 1993). This is difficult terrain for feminists, with the spectre of
essentialism menacing any consideration of embodiment in relation to
gender and sexual identity (Fuss, 1989). But its avoidance became increasingly problematic, giving rise to a growing realization that 'to separate the
feminine from female morphology is misguided theoretically and politically
even in strategic contexts' (Gross, 1986: 136). The concept of difference-in-
relation requires a 'theory of the flesh' (Moraga and Anzaldua, 1981: 23) to
elaborate an understanding of individual and collective becoming situated
in webs of connection that are practical as well as discursive; corporeal as
well as cognitive. Elizabeth Grosz's elaboration of a 'corporeal feminism'
(1989, 1994) provides perhaps the most sustained attempt to articulate just
such a theory.[7] She builds on Irigaray's understanding of difference as
always inscribed in/through the lived experiences of sexed bodies.

> I want to go back to the natural material which makes up our bodies, in
> which our lives and environment are grounded . . . a latent materiality
> which our so-called human theories . . . move away from [and] progress
> through . . . with a language which forgets the matter it designates and
> through which it speaks. (Irigaray, 1986, *Divine women*, quoted in
> Grosz, 1989: 172)

Here, the body is considered not as the passive container of logocentric
social being but as a living assemblage of corporeal dispositions and
relations which both register and orient our senses of the world. While
always configured through particular social orderings of meaning, technology and practice, these corporeal properties are no less conditional of the
very capacities of cognition and communication that mark the abstracted
ideals of individual autonomy and human distinctiveness. As Gail Weiss
(1999) goes on to suggest in more recent work, such a 'thinking through

the body' undermines the political myth of self-authorship and the privileged ethical status of humans as uniquely rational subjects, attending instead to the *inter*-corporeality of social conduct.

A second theme in feminist ethics that is particularly pertinent to the elaboration of an inter-subjective conception of the situated self is the praxis of care. This builds on the contention that feminisms can only move beyond 'the impasse of (in)difference' (Probyn, 1993) by simultaneously articulating questions of who am I with those of who is she? This ethical incarnation of difference-in-relation derives from a number of impulses in feminist work other than philosophy, particularly from psychoanalytic feminism (Meyers, 1994). A major stimulus to such work has been the empirically derived contention of psychologist Carol Gilligan (1982) that women tend to articulate more relational senses of self and stronger senses of responsibility for connected others than men – what she called a 'different ethical voice' to that institutionalized in conventional justice. The recognition and enacting of these relational senses of self and responsibility constitute what has become known as the 'feminine care ethic'. While the subject of dispute, this notion is concerned with ethical praxis and the corporeal register of connectivities which secure the well-being of those least mobile and most vulnerable, not as discursive subject-positions but as kindred mortals, such as the hungry, the sick and the abused (Lovibond, 1994). This understanding of ethical agency and community recognizes a bodily intentionality to human existence and social life that knits together multiple and apparently fragmentary collective identities, each of which is itself the outcome of a multiplicity of prior and present praxes (Kruks, 1995: 15).[8] Such an understanding certainly helps to substantiate an appreciation of inter-subjectivity in corporeal terms, but this has tended to be restricted in feminist accounts to relations between exclusively human subjects. As Vicki Kirkby suggests,

> it is so obvious that 'the subject' means, in fact, 'the human subject' that it goes without saying. . . . Even theoretically ambitious feminisms unwittingly tend to repair the sovereign subject through a politics of inclusion that would restore humanity its full identity. (1997: 151)

It is here, in expanding the corpus of 'beings that count', that environmental ethics promises to make an important contribution.

environmental ethics: enlarging the subject

> The multiplicity of living organisms retain, ultimately, their peculiar if ephemeral characters and identities but they are . . . mutually defining. (Callicott, 1989: 111)

In contrast to much feminist work, environmentalists have invested considerable energies in trying to extend the ethical domain of the autonomous self as a bearer of social rights beyond what Simon Glendinning calls 'the self-presence of human being' (1998: 9).[9] This has taken shape through the often disputatious ethical currencies of animal liberation and environmentalism (Luke, 1997). The first of these, which might be termed moral extensionism and is associated with longstanding concerns with animal rights, transports the liberal figure of the individual rights-bearing 'person' wholesale to a range of non-human creatures (see, for example, Cavalieri and Singer, 1993; Wise, 2000). These extensions are made either on the cognitive criteria of reasoning and linguistic capacities, which are usually restricted to primates and cetaceous mammals, or of sentience, a more inclusive criterion centred on the capacity to suffer or experience pain and covering all mammals with a central nervous system. Informed by new perspectives in animal biology and psychology, particularly primate cognition, this approach culminates in the proposal of a 'subject-of-life' criterion for extending ethical standing to all animate beings (Regan and Singer, 1989). Such approaches build on mainstream utilitarian or Kantian arguments and are open to the critiques of liberal individualism rehearsed above (see Benton, 1993), as well as the more far-reaching problems of trying to gauge the alterity of other animals by extending the humanist register of ethical considerability across the Cartesian divide.

The second approach, broadly aligned with deep ecological perspectives and sometimes informed by Gaian organiscism, has involved the elaboration of various notions of expanded human consciousness to encompass a recognition of our embeddedness in constitutive relations with the non-human world (for example, Macauley, 1996; Gottlieb, 1997). These efforts do not restrict the extension of ethical standing to non-human animals but include vegetal organisms, inanimate elements and even the planet itself under the collective term of 'earth others' (Bigwood, 1993). This enlarged ethical community frequently relies upon the invocation of a metaphysical dimension to being-in-the-world that sits uneasily with the triumph of reason associated with the humanist ideals of western political philosophy (see, for example, the critique of Ferry, 1992). Such efforts share the conviction that the human should not be the measure of ethical considerability and, just as importantly, that we cannot 'expect to have justice within the human community if we do not consider what this concept might mean in terms of our relations with nonhumans' (Gottlieb, 1997: xiv). The ironic conundrum of many such efforts, such as Mathews concept of the 'ecological self' (1991), Naess's notion of 'self actualisation' (1989) and Fox's idea of the 'transpersonal self' (1990), is how to square their insistence on the ethical standing of non-human nature in 'its own terms' with the ineluctable humanism of their non-anthropocentric enterprise. In a sustained critique of these approaches, Plumwood has identified them with what she calls the 'imperialism of the self' (1993), in which the ethical considerability of the non-human world is subsumed into the compass of human being, even as they strive to construct an inter-subjective conception of ethical agency. This highlights a key dilemma for environmental ethics. Feminist difficulties with the masterful standard of cognitive and linguistic competences, from which the ethical subject is fashioned, are amplified for environmentalists whose constituency of would-be subjects is more thoroughly excluded from this self-evident company than any human (Dryzek, 1990).

This dilemma has stimulated an important development in recent work on environmental ethics. Picking up Kruk's insistence on a materially situated, practically engaged, self as the embodiment of an inter-subjective understanding of ethical agency, this work has begun (re)exploring an understanding of relations between the self and the world centred on the corporeal immersion of humankind in the biosphere. This conceptualization of inter-subjectivity recognizes humans as 'beings thoroughly entwined with an extralinguistic world . . . [and that] to deny this entwinement is to bind ourselves to a quest for an abstract and empty sovereignty that destroys the world and is self-defeating' (Coles, 1993: 231). Like feminist evocations of a 'theory of the flesh', some of these explorations draw inspiration from traditions of dialectical reasoning, like that of Adorno (Coles, 1993), or phenomenology, such as the work of Merleau-Ponty (Abram, 1988; Langer, 1990). One of the most suggestive such parallels is that between Luce Irigaray's reworking of Heideggerian concerns with the neglected materiality of (human) being in *The forgetting of air* (1999) and David Abram's *The spell of the sensuous* (1997) in which he articulates the relationality of a living world through the same vital medium to conjure a

> . . . breathing landscape [that] is no longer just a passive backdrop against which human history unfolds, but a potentized field of intelligence in which our actions participate. As we awaken to the air, and to the multiplicitous others that are implicated, with us, in its generative depths, the shapes around us seem to . . . come alive. . . . (Abram, 1997: 260)

The emphasis in both these accounts is simultaneously on the corporeal embeddness of cognitive processes in the visceral dynamics of brain, eye and skin, etc., and the con-figuration of human well-being with and through that of other living beings (Matuarana and Varela, 1992). Arguably it has been at this intersection between feminist and environmentalist work that most has been achieved in terms of transforming these ideas into an ethical praxis (e.g. Curtin, 1991; Donovan, 1993).[10] Donna Haraway, for example, credits environmental feminisms with having been 'the most insistent on some version of the world as active subject, not as a resource to be mapped and appropriated by bourgeois, marxist or masculinist projects'

(1991a: 199). Variously translating the webs of connectivity between the life practices and well-being of different and particularly situated kinds, these practical configurations of an ecological care ethic begin by acknowledging that

> to be embodied is to be capable of being affected by the bodies of others and . . . is both a necessary and a sufficient condition for the generation of a bodily imperative [that] attend[s] morally to the needs of bodies who are unable to articulate those needs for themselves, the young, infirm, dehumanized and includes bodies that are not human. (Weiss, 1999: 162–3)

The feminist and environmentalist approaches to the subject of ethical considerability and community sketched in this section are ongoing and contested discourses which inform a wide variety of political practices, including on occasion each other, in ways that exceed the partiality of my rendition. The main contributions that I would attribute to the particular threads which I have traced are their various journeys towards a corporeal conception of ethical considerability and conduct that starts to engage the extra-linguistic 'forces, bonds and interactions' which Michel Serres urges on our attention in the passage which opens this chapter. Moreover, they are suggestive of the importance of spatial imaginaries and practices to challenging the myopic parameters of ethical connectivity in ways which complicate the geographies of intimacy and affect that configure conventional understandings of the proper extent of 'our' worldly responsibilities.

Equally, however, these approaches share shortcomings which are important in terms of my broader argument. Even amidst the talk of inter-subjectivity, embodiment and embeddedness, these accounts tend to treat the 'human' and 'non-human' in terms of 'social interactions between already constituted objects' (Rajchman, 2000: 12), thereby reiterating an a priori distinction between separate worlds in need of some kind of remedial re-connection. As a consequence, the remedies suggested by these feminist and environmentalist accounts, however inadvertently, retain a residual humanism that restricts the reconfiguration of ethical practice to terms in which the 'best the non-human can get out of [it] is to be permanently represented [by 'us'] as lesser humans' (Haraway, 1992: 86) while the subject of human-ness itself remains largely uninterrogated. Moreover, although the distinction between general and concrete others is a heuristic device which has no necessary spatial predisposition, feminist and environmentalist care ethics have tended in practice to map it simplistically on to the geographical binaries of distance/proximity, global/local, outside/inside, for example in the praxis of a maternalist model of care (Ruddick, 1989) or a localist model of ecology like 'bioregionalism' (Cheney, 1989). I now turn to consider the implications of taking hybridity seriously as a means of disrupting the residual hold of this purification of (human) society and (non-human) nature, and of the autonomous individual as the locus of ethical subjectivity, to begin to explore some alternative cartographies of/ for living in a more than human world that apprehend the fleshy currency, or what Gail Weiss (1999) calls the 'bodily imperative', of being-in-relation with and through heterogeneous others.

hybrid cartographies for a relational ethics

> What is inter-subjectivity between radically different kinds of sub-jects? How do we designate radical otherness at the heart of ethical relating? (Haraway, 1992: 89)

Bringing ideas of difference-in-relation to bear on the question of political and ethical community has been most extensively explored in the work of Haraway and Latour in their elaboration of concepts of hybridity. Har-away's argument is that we 'cannot not want' something called humanity because nobody is self-made, least of all humans (1992: 64). But in order to recuperate a progressive commitment to humanity as a moral community the dualisms associated with humanism have to be jettisoned. This line of argument informs several so-called 'post-human' efforts to reconfigure ethical competence and conduct by disturbing the consolidation of differ-ence at the borders between the 'human'/'non-human'. As Judith Halber-stam and Ira Livingston suggest,

> the human functions to domesticate and hierarchize difference within the human (whether according to race, class, gender) and to absolutize difference between the human and the nonhuman. The posthuman does not reduce difference-from-others to difference-from-self, but rather emerges in the pattern of resonance and interference between the two. (1995: 10)

Haraway's cyborg figure (1985), for example, articulates a political vision which appreciates the unstable and porous borders between human, animal and machine and the multiple modalities of subjugation that such an appreciation brings into view. Here, the possibilities of social agency are constituted through 'webs of connection' between radically different and particularly embodied subjects, connectivities that are fashioned through what she calls 'shared conversations' and 'semiotic-material technologies' (1991a: 192). Ethical praxis likewise emerges in the performance of multiple lived worlds, weaving threads of meaning and matter through the assemblage of mutually constituting subjects and patterns of association that compromise the distinction between the 'human' and the 'non-human'.

As with so many of Haraway's provocative ideas, what she means by 'semiotic-material technologies' is hard to fix. Her favourite examples are the body-technologies of prosthetics, genetics and organ transplants in which particular codified knowledges become stabilized as technological artefacts which, in turn, are grafted into and mobilized by living beings. These examples tend to site the dilemmas of hybrid subjectivity, and the cyborg figure used to signify them, within an individuated body-subject – 'a hybrid creature composed of organism and machine' (1991b: 1). There is a tension then in Haraway's account of the status and configuration of her hybrid subject – the cyborg. It is not clear whether, as Kruks asks, these hybrid subjects stitch their own parts together, in which case they become more cohesive than Haraway wants to admit, or whether this 'stitching together' is better understood as an operation taking place from without (1995: 9). If the first, then Haraway's hybrid subject falls back on an account of political and ethical agency which privileges cognitive and discursive faculties in the constitution of 'knowing selves' (however partial or unfinished the project of self-fabrication). If the second, then it is not clear from Haraway's account just what it is that connects these diverse knowing selves together other than the capacity for 'shared conversations'. As Vicki Kirkby observes

> Haraway's 'disassembled and reassembled recipe' for cyborg graftings is utterly dependent upon the calculus of one plus one, the logic wherein pre-existent identities are *then* conjoined and melded. The cyborg's chimerical complications are therefore never so promiscuous that its parts cannot be separated, even if only retrospectively. Put simply, for Haraway, there once was not a cyborg. (1997: 147, original emphasis)

I am not so sure that it is (ever) *that* simple for Haraway. But, while her account of hybridity successfully disrupts the purification of nature and society and the relegation of 'non-humans' to a world of objects, I agree that it is less help in trying to 'flesh out' the fabric of connectivity that transacts difference and, therein, the promise of a more than human ethical praxis. Such an exercise requires closer scrutiny of the inter-corporeal complications of heterogeneous life practices, or what Deleuze and Guattari characterize as the 'overlapping territories of affectivity and becoming' (1988: 267), than Haraway's cyborg figuration of hybridity seems to conjure.

In this context, I find Latour's account of hybridity, figured in terms of a 'net-working' effect, more suggestive for elaborating a relational understanding of ethical considerability and conduct. This networking ontology, like the rhizomatics of Deleuze and Guattari, emphasizes the affective relationships between heterogeneous actants, distributing their morphological particularity and mutability through all manner of energetic exchanges within and between them.[11] Cast in these terms, hybridity, signals not just the inter-connnectedness of pre-given entities but the condition of immanent potentiality that harbours the very possibility of their coming into being. Moreover, Latour spells out the difference that this interpretation of hybridity makes for the re-ordering of ethical community beyond the 'human'. Hybrid networks he argues force us to 'take into account the objects that are no more the arbitrary stakes of [human] desire alone than they are the simple receptacle of our mental categories' (1993: 117). Articulated through the cartography of networks (or rhizomes), hybridity disturbs the habits that reiterate the cumulative fault-lines between human/subjects and non-human/objects prescribed by an ethical reasoning abstracted from the particularity of embodiment and territorialized as the exclusive preserve of a 'Society' from which everything but the universal human subject has been expunged. Instead a multitude of affective actants-in-relation take and hold their shape performatively, as precarious achievements whose durability and reach is spun between the potencies and frailties of more than human kinds. It is in assemblages such as these that the 'forces, bonds and interactions' of Serres' cryptic 'natural contract' can make their presence felt in the vital topology of ethical relations; lived relations which are neither rooted to the spot nor the culmination of some singular chronology, but which stretch and fold multiple space–times through provisional alignments of polyvalent rhythms and passages of bodies and elements, energies and devices, memories and skills.

Latour's account of hybrid networking involves an important shift in tense from relational 'being' to relational 'becoming' and a more fluid sense of the spatiality and temporality of hybridity than Haraway's cyborg figure. These are important steps for my attempts to chart a topology of ethical relating. Latour's own gestures towards the ethical and political import of his account of hybridity bring us to his image of a 'parliament of things' (1993) or, as he has put it more recently, 'to the point that, today, the whole planet is engaged in the making of politcs, law and, soon I suspect, morality' (1999a: 214). But this is the point at which I find Latour's extraordinary work most problematic because of its apparent indifference to the witness of those living (and dying) at the sharp end of technoscientific re-orderings (see Star, 1991a). There is, as I suggested in chapter 3, an important divergence in analytical stance between ANT's emphasis on the effectivity of (quasi)objects and that of feminist science studies on the affectivity of (body)subjects. But there is something more than this. As Mark Elam has noted, there is something scriptural in the demeanour of Latour's writing that 'assumes a position outside of action, only to re-appear as science-in-action personified . . . [it is as though he] cannot help re-enacting the imperial ambitions that infuse the networks he charts' (1999: 21). In contrast, say, to Haraway's sustained commitment to 'taking sides' or even the 'cosmopolitics' of Isabelle Stengers which he so admires, Latour is too chary of situating his own knowledge practices or risking his intellectual acumen by association beyond the academy to nourish the kinds of connection between analytical adventure and everyday apprehension that are the measure of the 'passionate' mode of enquiry that I am after here.[12]

My argument throughout this book has been that it is both more interesting and more pressing to engage in a politics of hybridity that is not defined as/by academic disputes like the so-called 'science wars', important though these are, but in which the stakes are thoroughly and promiscuously distributed through the messy attachments, skills and intensities of differently embodied lives whose everyday conduct exceeds and perverts the designs of parliaments, corporations and laboratories. For those of us trained to earn our living in such centres of calculation, the epistemological imperative to acknowledge the situatedness (and affectiveness) of our own knowledge practices is at least well rehearsed, particularly in feminist science studies. Of more concern to me in this section of the book is how little onus science and technology studies seem to place on according such close and respectful analytical attention to the practical knowledges and vernaculars of everyday sense-making (Shotter, 1993). But this is no less vital if, as Haraway insists,

> taking responsibility for the social relations of science and technology means refusing an anti-science metaphysics, a demonology of technology, *and so* means embracing the skillful task of reconstructing the boundaries of daily life, in partial connection with others, in communication with all our parts. (1985: 100, my emphasis)

So, let me return to that most mundane of worldly transactions, eating, to illustrate the steps I have taken here towards a relational ethics that places corporeality and hybridity at its heart. As I noted in the introduction to this section, food is one of the most potent vectors of the 'bodily imperatives' that enmesh us in the material fabric and diverse company of 'livingness'. The skills and (dis)comforts of growing, provisioning, cooking and eating have long accommodated and intensified the proliferation of hybrids – through the cultivation of plants and animals; the wayward energies of wastes and additives circulating in water, soils and in the flesh; and the bacterial mutations and viral infections that traffic between life and death (Cone and Martin, 1997). The rhythms and motions of these inter-corporeal practices configure spaces of connectivity between more-than-human life worlds; topologies of intimacy and affectivity that confound conventional cartographies of distance and proximity, and local and global scales. These are the kinds of performative and immanent geographies of/for relational ethics that I have been working towards in this chapter; 'projects of making' more livable worlds made possible by the 'ongoing interweaving of our lives' with manifold others (Ingold, 2000b: 69).[13]

As I suggested in the previous chapter, food scares are events that condense the metabolic intimacies habitual to eating, mapping gut apprehensions into cogitable rationalities that are discordant with those of industrial food production. Perhaps the most archetypal such event in recent times is that known popularly as Mad Cow Disease. An 'unintended consequence' of the intensive feeding regime of the industrial cow, this disease took passage through protein meal supplements derived from rendered animal carcasses (including those of cows) and routinely fed to cattle (and other animals) to speed growth and increase bodily productivity. Its manifestation in an epidemic of the degenerative brain disease BSE (Bovine Spongiform Encephalopathy) in Britain's cattle population in the 1980s and 1990s turned out, against scientific and Government assurances at the time, not to stop there. Humans too began to exhibit pathologically similar and equally fatal symptoms of infectivity in cases numerous and distinctive enough to be categorized as a new variant of CJD (Creutzfeldt Jakob Disease) (see Ratzan, 1998).[14]

The ethical (and political) import of the BSE–vCJD epidemic in Britain begins by acknowledging the corporeal specificities of cows as herbivorous ruminants, and following the incongruous 'rationale' of a feeding regime indifferent to them through to the eating habits and food choices of consumers. The practice provoked revulsion and disbelief in equal measure among an unsuspecting public. What kind of rationality was it that could make sense of such routine cannibalism? The rationalities both exposed and overshadowed by the spectre of the disease were those of cost-cutting and profit-margins in a corporate animal feed industry careless of the offensive detail of how their products were derived, and of balance sheets and productivity gains for farmers accustomed to gauging their husbandry in terms of the metabolic conversion of inputs into outputs. At once 'man-made' and 'pathogenous', the hybrid potency of the disease resonated with gut apprehensions of the corporeal kinship and fleshy currency between cows and people. Mad Cow Disease became an uncanny familiar in homes and workplaces, conversations and mass media, extending its presence through all manner of intermediaries: the sombre figures and graphs plotting the rising incidence of disease and declining sales of meat (particularly beef); hidden camera journeys into the once alien worlds of abattoirs and rendering plants; and the sickening image that still endures of a cow staggering, collapsing and trembling on a concrete farmyard floor. Scientists, government ministers and industry spokespeople were disconcerted to find their authoritative pronouncements and scripted assurances enmeshed as more or less compelling storylines in this intricate national drama (Miller, 1999).

The symptoms being so widely witnessed only became officially consolidated around the acronym BSE in 1987. Early government policy towards the disease was informed by the report of the Southwood Report (1989) which broadly accepted the then preferred theory that scrapie, a disease endemic in the sheep population, was the most likely infective agent being transmitted through animal feed to cattle. The Southwood Report concluded that 'from the present evidence it is most unlikely that BSE will have any implications for human health' (ibid: conclusions para. 9.2). Seven years later in 1996, the British Government finally admitted that the disease could, and had, spread to humans. This about-face followed the unwelcome advice of its Expert Panel on BSE in 1995. The Panel accepted evidential claims supporting another theory, that the infective agents in BSE (like scrapie and other transmissable spongiform encephalopathies – TSEs) were 'proteinaceous infectious particles', or 'prions' (see Ridley and Baker, 1998). They concluded that it was much more likely that the BSE epidemic had been caused by the recycling of BSE-infected cow carcasses in cattle feed, rather than those of scrapie-infected sheep, and that the disease was capable of transmission to humans through the ingestion of infected tissue and body fluids, blood transfusion and the like.[15]

As the sticky properties of prions gained scientific and policy adherents, so too did they become potent spokes*things* for the porosity of the corporeal borders between cows and people, effecting their indifference to species location and slow tempo replication in the spatial and temporal ordering of agri-food networks (see Hinchliffe, 2001). As such, they bore credible witness to the metabolic geographies bodying forth in the gut apprehensions of eating. Incarnating connectivities between the sites and practices of food production and consumption, animal and human well-being, these 'rogue' proteins proved unlikely allies in undermining the prevailing commercial, policy and scientific cartographies of affectivity and responsibility and making space for more relational ethical possibilities. The realignments of inter-corporeal sensibilities to collective modes of sense-making nourished by BSE have been honed through any number of subsequent food scares in Europe, galvanizing changes in shopping and eating habits, producing and retailing practices, and policy architectures and instruments. Red meat, particularly beef, consumption has never recovered. More and more people are choosing organic and/or animal welfare certificated foods. The rationale and practices of product traceability developed in these 'alternative' food networks, are increasingly becoming mainstream marketing and policy standards (see Murdoch et al., 2000). Such polyvalent mappings into knowledge of the affectivity of embodied difference diagram new modes of connectivity that are the stuff of ethical relating.

becoming other-wise

> . . . in the cyborg context of . . . hybrids of nature/culture, the
> question is not who will get to be human, but what kinds of couplings
> across the humanist divide are possible – or unavoidable. (Wolfe,
> 1998: 84)

In an effort to articulate a relational understanding of ethical connectivity
that does not presume or reinforce the cartographies of humanism, I have
identified corporeality and hybridity as key modalities for reconfiguring the
spaces and constituencies of ethical practice. Far from abandoning the
collective moral claims of humanity, this enterprise is concerned with
recuperating them from the grip of a universal ethical subject configured as
the autonomous self, and recognizing that their efficacy depends on
admitting more than human difference into the compass of considerability.
As Katherine Hayles argues in her account of *How we became post-
human*,

> to think of the subject as an autonomous self . . . authorizes the fear that
> if the boundaries are breached at all, there will be nothing to stop the
> self's complete dissolution. . . . When the human is seen as part of a
> distributed system, the full expression of human capability can be seen
> precisely to *depend* on the splice rather than being imperiled by it. (1999:
> 290, original emphasis)

But, like the traces of one-plus-one logic that haunt Haraway's cyborg, the
'splice' here betrays Hayles's post-human as a cybernetic novelty; an
epochal breach in the otherwise settled borders between the human and the
non-human, and one expressive of 'our' capabilities. The splice merely
stitches over the cut. The kind of relational ethics that I have been working
towards here casts hybridity in a different tense, defined less by its
departure from patterns of being that went before than with how it
articulates the fluxes of becoming that complicate the spacings–timings of
social life, and expressive of the creative impulse of more than human
energies.[16] On this account, hybridity compels us to acknowledge that not
only does 'humanity' always already 'dwell among badly analysed composites [like nature or the non-human] but that 'we' ourselves [the human-all-
too-human] are badly analysed composites' (Ansell-Pearson, 1997: 7).

Taking feminist and environmentalist critiques of the individualist
currency of mainstream ethical discourse as my starting points, I have
argued that their various efforts to articulate more relational ethical praxes
by either 'embodying' or 'enlarging' the company of ethical subjects are
often thwarted by a residual humanism that condemns them to trafficking
between (human)/society and (non-human)/nature as pre-constituted
domains of categorically different kinds of being. The radical pluralism of

hybridity variously invited by science studies can only do its work by refusing the Cartesian terms of this settlement in which 'human identity is wagered entirely on the use of "words", while the animal body, with all its inarticulate sounds, is relegated to the mechanical universe of automatons and chiming clocks' (Senior, 1997: 62). It is a settlement that both diminishes human conduct, reducing it to the dictates of a disembodied reason, and disqualifies everything else from the company of agential efficacy. In so far as they temper the lingering one-plus-one calculus of 'couplings', hybridity and corporeality redirect our attention to the affective relations between heterogeneous bodies in terms of their specific enunciative consistencies within a material–semiotic register of mutual prehensions and sensibilities that exceeds the signal monopoly of the word (Hansen, 2000a: 13). Learning (how) to map these affectivities into knowledge, like

> learning to swim or learning a foreign language means composing the singular points of one's own body or one's own language with those of another shape or element which tears us apart but also propels us into a hitherto unknown . . . world of problems. (Deleuze, 1968: 164)

In the manner of 'food scares', hybridity and corporeality trip those habits of thought that hold 'the body' apart from other bodies and 'the human' apart from other mortals, motioning instead to the shifting fabric of differentiation produced through their lively enfolding and which, as de Landa puts it, 'keeps the world from closing' (1999, 36). In this they amplify the repertoire of skills and associations enjoined in the praxis of ethical relating and help to open up the 'possibility and actuality of connections, arrangements, lineages, machines' (Grosz, 1994: 197) in at least three ways. First, by dispersing ethical considerability beyond the unified (and always) human subject without resorting to its wholesale extension to other living kinds. It is no longer, as Wolfe puts it, a 'question of who will get to be human' but rather one of how the 'we' of ethical communities is to be renegotiated on account of its heterogeneous, intercorporeal composition. Secondly, by complicating this bodily redistribution of ethical subjectivity in terms of the profusion of intermediaries – instruments, signals, machines, elements – which insinuate their energies and inertias in the intimate assemblages of corporeal becoming. Hybridity and corporeality interfere, in other words, with ethical prescriptions that would disown such familiars, making it possible

> instead of demonizing technologies [to] assess their promise and those of the new bodily configurations [afforded] by them in terms of the extent to which they promote and preserve the space of differentiation that makes our corporeal exchanges possible. (Weiss, 1999: 6)

And thirdly, by releasing the spatial imaginaries of ethical community from both the geo-metrics of universalism and the confines of propinquity and genealogy, they disturb the territorializations of self, kinship, neighbourhood and nation and invite other 'languages of attachment' (Ignatieff, 1984: 139).

Notes

1 For example, environmental sociology (Hannigan, 1995); environmental anthropology (Descola and Palsson, 1996); environmental history (Bird, 1987); and environmental politics (Dobson and Lucardie, 1995).

2 Of course there are notable exceptions to this generalization. Massey (1999b) provides a useful survey of, and intervention in, these ongoing conversations.

3 There are remarkable echoes in Fitzsimmons's critique of Marxist geographers' 'peculiar silence on the question of nature' of Merleau-Ponty's critique of the 'astounding' lack of attention of Marxist philosophers to the question of nature (1970: 63). Still the most important exception to this 'silence' identified by Fitzsimmons is Neil Smith's book, *Uneven development* (1990/1984).

4 For a critique of dialectical analysis within the geographical fold, see Castree's critique (1996) of Harvey's treatment of nature (1996). On a larger canvass, Cary Wolfe's *Critical environments* explores the problem for Marxist dialectics in terms of the dilemmas of retaining a strong sense of contradiction without it degenerating into 'mere antinomy' or falling into the false assurance of some notion of teleological inevitability (1998: 132).

5 The significance of this stance is that it unsettles any account which is inclined to render messy fragile net-workings as slick consolidated totalities like Science, Capitalism, or the State and, so, recovers a myriad of life-size orderings overshadowed by their heroics (see Gibson-Graham, 1996). The description of such ontological stances as 'modest' is first, and best, made by John Law (1994), but see also Callon and Latour (1981). The term 'situated knowledges' is, of course, Donna Haraway's (1991a), but it has been widely misconceived as an argument about localizing positionality (for useful clarification, see also Haraway, 2000: 71).

6 While publications do not capture the half of these conversational networks some useful way-markers include review essays (Murdoch, 1997a, 1997b); a retrospective on ANT by 'practitioners' (Law and Hassard, 1999); and two journal special issues published in 2000 – *Society and Space* (18/2) on ANT (edited by Hetherington and Law) and *Body and Society* (6/3–4) on 'bodies of nature' (edited by Macnaghten and Urry). See also Michael (2000).

7 I have some problems with amalgamating the diverse countercurrents to the humanist assumptions of social theory into a singular notion like Nigel Thrift's 'non-representational theory' because of the irresistable tendency for 'it' to be

reified along the way – in the manner of ANT. However, in so far as it provides a serviceable 'flag of convenience' that fosters conversations and alliances between diverse theoretical projects and impulses that variously challenge 'I think before I act' conceptions of social agency and freight more radical means to register the heterogeneity of social life then I am content to sail under it.

8   The net-workings of ANT are reminiscent of the 'rhizomatics' of Deleuze and Guattari (1988), a connection on which Latour comments directly in an interview with Crawford (1993: 262).

9   As the term bio-philosophy implies, the contributions of biologists and philosophers are thoroughly mixed in this enterprise. For example, contemporary biologists like Margulis and Fester (1991), Matuarana and Varela (1992), Goodwin (1994) and Rose (1997) have made striking contributions to these debates.

10   The cognitive and linguistic competences that conventionally define the fully-fledged subject and social actor are patriarchal constructs from which various categories of 'humans' have been, and are still being, excluded. Moreover, their status as the distinguishing mark of 'humanity' is troubled by the comparable skills of other classes of animals (notably, primates and cetaceous mammals) and broader reassessments of animal cognition (see Ingold, 1988b; Noske, 1989).

11   Merleau-Ponty died working on a new philosophy of nature that was elaborating an 'account of Earth as an intertwining (after Husserl) and enfolding of humans within nature that is an embrace' (Johnson and Smith (1990: xxxi). Some preliminary sense of this project can be gleaned from the collection of his lectures (1952–60) in which 'nature' is one of his themes (Merleau-Ponty, 1970).

12   Some years after graduating from UCL, I returned as a research assistant and so crossed the threshold of the staff common room through a door with the inscrutable nameplate 'Maconochie Room'. It was here that I first encountered Australia. Overseeing the daily rhythms of coffee, sandwiches and talk was, and still is so far as I'm aware, a short rank of monochrome (male) figures, commemorating the passage of eminent professors. Most celebrated among them was Captain Alexander Maconochie, first Professor of geography at UCL in the 1830s, Secretary of the Royal Geographical Society in 1833 and one-time Governor of the penal colony on Van Dieman's Land (now Tasmania) in the 1840s, but dismissed for his pursuit of a reformist agenda (Clay, 2001). The very same. In that 'first encounter' cartographies of empire and discipline collided with my own itinerant childhood spent among various of Britain's faded military outposts, tripping over the unspoken colour-coded cordons of daily life.

Section 1: Introduction

1   A series of variously negative responses to Cronon's essay were published in a special issue of *Environmental History* in 1997. Both the temper and substance of these responses attest to the deeply-rooted place of the wilderness ideal in

the institutional fabric and popular culture of North America, particularly the United States, that Cronon sought to expose as dangerous 'habits of thinking' (1995: 81).

2   See also the interventions of the Luke's *Eco-politics* (1997) from a North American (US) perspective, and Ferry's *Le nouvel ordre écologique* (1992) from a European (French) perspective.

3   The exception is a rather fragile line of interest in domestication that links, in very different ways, the work of Geographers like Ian Simmons (1996) and Robin Donkin (1989) and more recent post-colonial cultural geographies (e.g. Anderson K., 1997; Willems-Braun, 1997) with the cultural geography of Carl Sauer (see, Leighly, 1963) and the bio-geography of the 1950s and 1960s (e.g. George, 1962).

chapter 2

1   Ridley Scott's blockbuster movie *Gladiators*, released a year or so after an earlier version of this chapter was published, provides a telling reminder of the enduring fascination of the Roman gladiatorial spectacle in western popular culture. The camera works to incorporate the cinema audience into the spectatorial ranks of those in the amphitheatre, seducing them/us into sharing the visceral passions of the event.

2   Modern biology, informed by cellular ultrastructure through electron micros-copy and detailed knowledge of gene sequences, is rewriting these long-standing classificatory systems (Sagan, 1992). See, for example, Margulis and Schwartz (1982) and Woese *et al.* (1990).

3   While species interactions have become the credible subject of Ecology, and the social dynamics of animal groups have attracted some passionate exponents, the funding priorities and professional culture of the biological sciences remain wedded to a Cartesian conception of animal life as enumerable biological units of interest primarily in terms of their aggregate population trends or genetic diversity (Senior, 1997).

4   Interestingly, the taxonomy of such cat-like creatures in the modern zoological science that superseded the vagaries of this Roman nomenclature is itself under assault from genomic classificatory methods (see note 2 above). Recent evi-dence presented at the American Genetics Association, for example, has suggested that the phenotypic menagerie of 32 sub-species of Puma (including panthers and cougars) compromises only six genetically distinguishable sub-species (Pennisi, 1999: 2081).

5   As Jennison points out, some degree of crowd-pleasing poetic licence seems likely to be associated with the phrase 'African beasts', particularly in the Augustan age and later, making it less than reliable evidence of precise geographic origins (1937: 53).

6   Toynbee (1973) makes clear that such 'realistic' depictions of animal hunts and *venationes* have to be seen in the context of a much larger repertoire of animal art, particularly mythological and pastoral renditions but also commemorative depictions of companion animals among the Roman elite, especially horses and dogs.

7   Grove (1995: 20) notes that by the reign of Emperor Claudius (AD 41–54) the river Tiber had silted up to such an extent that most commercial port activities had to be shifted from Ostia to Civitavecchia.

8   There are parallels here with the earnest turn to science in the debate about banning stag hunting in the UK to determine whether or not these animals 'really' experience trauma and fear in the chase and at bay by proxy measurements of hormonal and chemical levels in their bodies (Bateson, 1997). This is a classic example of what Ted Porter (1995) calls the 'pursuit of objectivity' in science and public life.

9   See Doyle (1997: 28) for an engaging example of this kind of slippage in the excitable language of genetic engineering involving the metonymic substitution of 'code-script' for organism in Schrodinger's treatment of the 'Chromosone'. It is a slippage much in evidence in the hyperbole surrounding the USA/UK science establishment's launch of the 'draft blueprint of the Human Genome' (*Guardian* supplement, 26/6/2000).

chapter 3

1   I accept that this technical tendency, or what Latour calls 'double-click' networks (1999b: 20), is a rather hackneyed translation of ANT but, nonetheless, the socio-technical emphasis is real enough in the corpus of work that pays allegiance to this potent acronym.

2   It is not that those working with 'ANT' have ignored non-human life forms, one has only to think of Callon's scallops (1986), Latour's microbes (1988) or even elephants themselves (Cussins, 1997), but rather to note the striking contrast in the kinds of non-humans that figure overwhelmingly in ANT accounts as against those in feminist science studies.

3   A sub-species of *Loxodonta africana* – *L. a. cyclotis* – has been described in the forests of west Africa, smaller in size than its more numerous savannah relatives (Kingdon, 1997).

4   Whether by accident or design the ISIS acronym mimics the name of the ancient Egyptian goddess of fertility.

5   Thus, for example, the accidental death of a keeper in the elephant enclosure at London Zoo in 2001, led to the re-location of the elephants to a wildlife park and the closure of one of the zoo's most popular visitor attractions.

6   The research at Paignton Zoo which informs this chapter coincided with the filming of a six-part BBC documentary series on the working life of zookeepers which was first broadcast in the spring of 1998.

7   Earthwatch projects span archaeology, palaeontology, geology and anthropology as well as life sciences. But the majority of field projects, about 87 of the 127 projects in the 1997 brochure and some 63 per cent of the total number of volunteers concern animal, plant, bird or marine life, and ecology (Scientific Officer, interview 10/11/97).

8   The appearance of Earthwatch on a special edition of the *Holiday* travel programme on BBC1 (16/11/99), with television personality Charlie Dimmock, suggests that this ambivalence remains salient even after the relaunch of Earthwatch as an Institute.

9   For a fuller exposition of this medical research, see Berg and Mol (1998).

Section 2: Introduction

1 This latest rupture in the legal fabric of property is often traced to a battle in the US courts in the late 1970s between John Moore, asserting rights to property in his person, and the medical centre at UCLA, asserting rights to intellectual property in the 'Mo cell line' derived (without his consent) from his spleen. The medics won (see Rabinow, 1992b). The Human Genome Project affords a more recent and concerted example of this 'vampire' mode of bio-prospecting (see Cunningham, 1998), stimulating the US Patent and Trademark Office to issue some 1,250 patents on human gene sequences in the last two decades of the twentieth century (Anderson L., 1999: 79).

2 This insistence is shared but differently framed and pursued in liberal (e.g. Reeve, 1986), Marxist (e.g. Tribe, 1978) and post-structuralist (e.g. Kelley 1990) accounts (for a discussion, see Shapiro I., 1991).

chapter 4

1 For example, Paul Carter (1987) cites the journals of Captain James Cook and the botanist Sir Joseph Banks about their encounters with Australia aboard the *Endeavour* (Banks Sir J., *The Endeavour journal 1768–1771*, J.C. Beaglehole (ed.), 1962. Sydney; and Cook J., *The journals of Captain James Cook on his voyages of discovery*, Volume 1: 'The voyage of the *Endeavour*, 1768–1771', J.C. Beaglehole *et al.* (eds), 1955. Cambridge). Ross Gibson (1992) cites the letters of early colonists like Thomas Watling which first appeared in his native Scotland in 1794, *Letters from an exile at Botany Bay to his aunt in Dumfries* and is now available in Foss P. (ed.), 1988. *Island in the stream*. Pluto Press, Sydney). William Lines (1991) cites the journals of Surveyor Generals Charles Sturt and Thomas Mitchell whose explorations of the Murray basin in the 1820s and 1830s recorded daily encounters with Aboriginal people, sometimes upwards of 200 in number, while still claiming to discover an uninhabited country (Sturt C., 1834 (2nd edition). *Two expeditons into the interior of southern Australia*. 2 volumes. Smith Elder & Co., London); and Mitchell T.L., 1839. *Three expeditions into the interior of Eastern Australia*. 3 volyumes, vol. 2. T & W. Boone).

2 These shifting renditions of the 'state of nature' and its inhabitants 'natural man' were by no means uncontested. Secular natural law theories met constant challenges from various currents of dissenting Christian moral discourse, from the Spanish Thomists during the fifteenth and sixteenth century Spanish colonization of Latin America (see Pagden, 1987) through to the philanthropic crusades against slavery and the abuse of Aboriginal peoples in late eighteenth- and nineteenth-century England. These impulses were perfectly compatible with colonialism as the same moral discourses underwrote the promulgation of Christianity as a 'civilizing' force, but by the end of the nineteenth century their currency had been decisively overshadowed by that of science in framing the project of empire.

3 See, for example, Brennan (1991); Pearson (1993b); Rowse (1993a); Dodson (1994); Reynolds (1996).

4 See in particular, Jane Jacobs' *Edge of empire* (1995) and more recent work with Ken Gelder *Uncanny Australia* (1998).

5 For this reason, and despite several references to Deleuze and Guattari's 'A thousand plateaus', it seems to me that Carter's spatial history bears rather more superficial resemblance to their 'rhizomatics' than some have claimed (see Rodman, 1993).

6 The British House of Commons Select Committee was established in 1834 following a motion by Mr Thomas Fowell Buxton, the political heir to the slavery abolitionist William Wilberforce and co-founder of the British and Foreign Aborigines Protection Society in the same year, who chaired its proceedings. The Committee reported in 1837 and, while Australia is not its primary focus, the hasty parliamentary passage of the 'South Australia Coloni- zation Bill' in the summer of 1834 provided a fresh stimulus to their con- demnation of colonial 'settlement' practices (Minutes of Parliament (GB), 1834). Among those called to give evidence to the Committee was a prepon- derance of clergymen, many of whom held out 'Christian instruction' as the only realistic recompense for the ill-treatment of Aboriginal peoples in the colonies.

7 Property preoccupied debates in liberal philosophy and political economy in the eighteenth and nineteenth centuries. Liberal currents took their lead from Locke, for whom private property is the cornerstone of civic governance and the liberty of the individual. Critics, from Marx and Engels to Henry George, cast it as the bastion of capitalism and/or class privilege. For useful overviews of these variegated debates, see Honoré (1961); Macpherson (1978); and Ryan (1984).

8 The familiar face of these heroic accounts of private property is the figure C.B. Macpherson identified as the 'possessive individual' (1962). Here, rights to occupy, use, alienate and benefit from land are bundled together and vested in a single person/proprietor as freehold or fee simple title (see also Vogel, 1988). But as several early modern historians and political philosophers have noted, this modern figure is not one that would have been recognized by those, like Locke and others writing in the natural law tradition, in whose texts it is now routinely originated (see, for example, Squadrito, 1979; Tully, 1993: chapter 2; Macfarlane, 1998).

9 This understanding of English Common Law was itself considerably more formalized than in earlier configurations of governance. The seventeenth- century jurist Sir John Davies, for example, observed in 1674 that

> The common law of England is nothing else but the Common Custom of the Realm. . . . It can be recorded and registered no-where but in the memory of the people. For a Custom taketh beginning and groweth to perfection . . . when a reasonable act once done is found to be good and beneficial to the people, and agreeable to their nature and disposition, then do they use it and practise it again and again, and so by often reiteration time out of mind, it obtaineth the force of a Law. (cited in Reynolds, 1996: 75)

10 Common law occupancy was key to the dispossession of customary or common land use in Britain (and Ireland), which had contributed a major component of the diet and fuel of the labouring poor (Goodrich, 1991). It was progressively removed by parliamentary statute – with some 100 Enclosure Acts between 1800 and 1834 (Neeson, 1993). Those dispossessed by these acts and/or who actively opposed them made up a considerable part of those transported as convicts and emigrants to Australia (Thompson, 1991).

11 While Blackstone's *Commentaries on the laws of England* met with a variable reception at home, they attained a substantial reputation in the colonies. Eight editions were published in his lifetime (1723–80) and another 15 by 1854 (Jones, 1972).

12 While the Mabo judgment's admission of 'native title' into common law broke new legal ground by establishing land rights without the jurisdiction of Parliament, it had been preceded by several piecemeal legislative provisions, notably under the Aboriginal Land Rights Act (Northern Territories) 1976 and in association with National Park designations. However, even where native title recognized at common law approaches 'full ownership' in the terms set out by the Mabo ruling, it remains subject to important limitations (see Bartlett, 1993: 42):

> i.   it cannot be alienated (or sold) other than by surrender to the Crown;
> ii. beneficial title is communal and held in trust while personal titles, whether of individual, family or band, are transferable by custom;
> iii. it does not constitute permissive occupancy and can be legally and validly extinguished by the Crown under strict conditions.

13 In this, the majority judgment accords with Kant's objection to natural law theorists like Locke or, later, utilitarian theorists like Bentham. He argued that while cultivation can confirm title, in that it signals to others that a piece of land has already been taken in rightful possession, this fact may be conveyed 'by many other signs that cost less trouble' (Kant, 1887: 97, cited in Vogel, 1988).

14 Indeed the majority judges' insistence that their ruling on the common law status of native title applied to the whole of Australia was based on a recognition that the circumstances of the Meriam case, characterized by the cultivation of garden plots passed through family lines between generations, did not set it apart in legal terms from those on mainland characterized by more seasonal patterns of land use and more collective proprietorial practices (see Rowse (1993b) for a discussion of the implications of the Mabo ruling for interpreting indigenous 'traditions').

15 Western Australia, Northern Territories and Queensland, the States with the largest proportion of their land under pastoral and mining leases and the highest proportion of indigenous Australians within them, were the most vigorous in their opposition to the Native Title Bill. They framed their opposition in terms of resisting efforts by the Commonwealth Government in Canberra to infringe their established regional jurisdiction over land management.

16 Two currents in these political wheeling and dealings are of particular note. In contrast to the sustained opposition of the corporate mining lobby, the Labour Government won the support of the National Farmers' Federation for their Native Title legislation, against the grain of many of their regional constituencies and their traditional allegiance to the National Party. As the Government was building these alliances, Opposition politicians led by the Premier of Western Australia, Richard Court, were engaging in political manoeuvres to undermine the Commonwealth legislative process by passing State legislation that extinguished common law native title and replaced it with an impoverished form of statutory provision.

17 There are occasional Opposition voices, notably that of Dr Wooldridge, Deputy Leader of the Liberal Party, whose speeches (Hansard (HoR), 24/11/93: 3599–603) are clearly sympathetic to native title and regretful of the lack of a bi-partisan approach to the legislation. But even these speakers follow their party whip to vote against the Bill.

18 Pauline Hanson lost her parliamentary seat in the 1998 elections and, at the time of writing, her One Nation party was being investigated by the Australian Police for fraud.

19 There are currently some 145,000 pastoralists, in comparison with 260,000 indigenous Australians, a number that has been in decline since the end of the First World War and is now falling at an accelerated rate. While a minority of pastoralists are among the wealthiest landholders in the country, corporate business forms are much less widespread in the pastoral industry than in the mining industry and the majority of them make moderate and uncertain livings (see Lawrence, 1990).

20 The Commonwealth Constitution makes only two passing references to Aboriginal people. One excludes them from the law-making powers of the Commonwealth (section 51 subsection xxvi) and the other excludes them from the census enumeration of the Commonwealth and the States (section 127), both of which were conceived in terms of governing Australia's immigrant population. For a discussion of the political context of the 1967 referendum see Attwood and Markus (1999).

21 In establishing their credentials for talking about/for Aboriginal people, MPs referred variously to the size of the Aboriginal population in their constituency; recollections of visits and interactions with Aboriginal communities; and, somewhat bizarrely, their appreciation of 'Aboriginal culture'. Tim Fischer (Leader of the National Party), for example, countered charges of racism against a party conference address he made by referring to 'my praise of Aboriginal culture as manifested through their rock art, crafts, music, dance and their tribal customs and practices' (Hansard (HoR), 23/11/93: 3425).

22 In addition to establishing a quasi-judicial regime for dealing with native title claims, past and future, the Native Title Act 1993 also set up a national fund for the acquisition of land by indigenous Australians who did not benefit directly from native title and foreshadowed a wider-ranging 'social justice package' (Peterson and Sanders, 1998: 21). These procedures and bureaucracies built on those already established under the 1967 Land Acts.

23 Senator Panzinni described himself as a 'first generation Australian' and noted that his own grandchildren would likely inherit only a 'fraction of his Italian blood'. The same notion of race as a biologically derived category was used by Michael Cobb MP (NSW, National Party) to question the credentials of Eddie Mabo as a plaintiff for the Meriam people whose case made history in the High Court, on the grounds that he was 'only' the adoptive son of a community family.

24 The association between moral order and settled social forms goes back to ancient Greece (see Greenblatt, 1991: 68–70). Its reconfiguration through the assemblage of private property was woven through a discourse of 'improvement' in which both the person and the land were honed through the discipline of labour. For Locke, for example, the justification of private property could not be exceeded by the size of a person's landholding except by 'the perishing

of anything uselessly in it' (1988 (1690), para. 46: 300). In similar vein, J.S. Mill declared two centuries later that

> Whenever, in any country, the proprietor, generally speaking, ceases to be the improver, political economy has nothing to say in defence of private property. . . . In no sound theory of private property was it ever contemplated that the proprietor of land should be merely a sinecurist quartered on it. (1961/1870: 231)

25 The Deputy Leader of the Opposition in the Senate, Senator Richard Alston (Victoria, Liberal Party), for example, declared the High Court ruling 'the greatest *obiter dicta* in history' (Hansard (Senate), 16/12/93: 5024). More explicitly, the Pastoralists and Graziers' Association of Western Australia, in their evidence to the Legal and Constitutional Affairs Committee on the Native Title Bill, argued that it was 'conferring a title proved for a settled agrarian and fishing community on a nomadic people of a completely different race and lifestyle' (quoted by Senator Chris Evans (Senate Leader of the Labour Party), Hansard (Senate) 14/12/93: 4583).

26 Reynolds (see particularly, 1988, 1992) interprets the Committee's conclusions, and the subsequent efforts by colonial administrators to govern the acquisition of land and treatment of Aboriginal peoples by settlers in Australia from Britain, as evidence of an overriding concern with protecting native rights and welfare. While this was unquestionably a priority in their considerations, the frequent contradiction between these efforts, the terms of the statutory Acts establishing colonies and what was happening on the ground, suggests that any such concerns were compromised in practice by commercial and fiscal considerations. Moreover the Christian morality which pervades the report, as noted above, was also a mainstay of the civilizing pretensions of colonialism.

27 Opposition leaders in the Commonwealth and some State Parliaments (notably those in Western Australia and Northern Territories) launched a concerted media campaign against Native Title, insinuating their territorial anxieties into the 'suburban backyard'. For example, John Howard, in his infamous 'One Australia' speech (17/11/93), warned that 'other Australians want to be sure that they do indeed own their own home or farm and that they won't have to go to court to defend them' (cited by Hon. Gary Johns (Queensland, Labour Party) (Hansard (HoR), 24/11/93: 3596). Similar scare tactics were mobilized in campaigns run by the corporate mining lobby, the Australian Mining Industry Council, and its spokespeople like Hugh Morgan (1992). Here cartographic propaganda, exaggerating the extent of Native Title on maps like that in figure 4.3, was a favourite weapon (see Gelder and Jacobs, 1998: 139–41).

28 These 'amendments' to native title have attracted an 'early warning' decision from the UN's Committee on the Elimination of Racial Discrimination, the first ever issued to a developed nation (Mercer, 1997).

chapter 5

1 The Convention on Biological Diversity has a much larger remit than domesticated plants and animals and takes its impetus from the conservation and

utilization of all living resources (see Takacs, 1996), relegating the question of Plant Genetic Resources (PGR) for agriculture to chapter 14 (CBD, 1992).

2   RiceTec Inc. of Texas applied to register Basmati as a company trademark in 1998. It later withdrew this blanket patent application in the face of concerted opposition from the Governments of India and Pakistan and other agri-technology companies in the USA. The decision of the US Patent and Trade-mark Office in August 2001 permitted the patent registration of three specific hybrid varieties – Texmati, Jasmati and Kasmati – developed by the company over a ten year period of cross-breeding with American long grain rice varieties. The decision was greeted by both the company and its opponents as a 'victory' (*The Guardian*, 23/08/01).

3   Following the request of Conference resolution (9/83), the Commission for Plant Genetic Resources (CPGR) was formally established within the terms of Article VI paragraph I of the FAO constitution by a resolution of the FAO Council (1/85) at its meeting of 24 November 1983.

4   The Group of 77 was an influential alignment of lately independent nations pursuing a New International Economic Order against the hegemony of industrialized countries, particularly the USA in the 1970s and early 1980s (Larschan and Brennan, 1983; Fowler and Mooney, 1990: 187–200). In 1975 it comprised 103 out of a total 138 member states of the United Nations. Its influence has dissipated since the late 1980s, replaced by the Cairns group alliance associated with World Trade Organization/GATT negotiations, and the strengthening of regional trading blocs like NAFTA (North American Free Trade Area), the European Union and ASEAN (Association of South East Asian Nations).

5   While the United States delegation was the most vociferous opponent of the International Undertaking, other industrialized countries, including the UK, Canada, France, Germany, Switzerland and Japan, shared some of its objec-tions. By 1993 the USA, Canada and Japan remain notable for their absence from the list of adherents to the Undertaking.

6   The Deleuzian terminology of the 'event' (see Deleuze, 1990/1969) that affects this analytical modality of immanence can be found elsewhere in bio-philosophy, for example, Whitehead's notion of 'concrescences' (1929), and de Landa's notion of 'temporary coagulations' (1997: 104) which is directly derivative of Deleuze.

7   These documentary archives are made up of written records from three main procedural bodies conducting the business of the FAO in relation to PGR. First, the official reports of the bi-annual Conference of the FAO, at which delegates from all member countries debate and vote on the business of the meeting. These reports minute the items of business, wording of resolutions and voting procedures and are identified by the documentary prefix – C (year/rep). Secondly, the verbatim records of Conference debates that are conducted through two Commissions charged with reporting back to the plenary sessions for voting. These records are identified by the documentary prefix – C (year/commission no./PV [proceedings verbatim]). And thirdly, the reports of the bi-annual meetings of the CPGR itself which record the agenda, discussion papers and minutes in which representatives of any FAO member country can participate. These are identified by the documentary prefix – CPGR (year/rep). Occasional reference is made to documentary reports and verbatim records of meetings of the FAO Council that is made up of about 50 representatives

elected by member countries and meets two or three times between Conferences. These are identified by the documentary prefix – CL (year/rep).

8 These slippery distinctions are elaborated for discussion as an agenda item at
the 2nd session of the CPGR. Wild relatives are identified as 'the products of
nature'; weedy relatives as the 'botanical bridge between wild relatives and
modern plant varieties'; primitive cultivars as 'plants that have evolved as a
result of both natural and human selection'; and 'modern varieties' as being
'the result of plant breeding' (CPGR, 1987a: 2).

9 It is worth noting that Weismann's theory of 'germplasm' (1892), which held
that the development of hereditary traits was a function of the reproductive or
germ-cells sequestered from the rest of the body (and its living environment),
derived from his experimental work with insect embryos and roundworms. As
developmental biologists like Bonner (1974) pointed out some time ago, this
'principle' does not hold good for plants (the cells of which are totipotent).
More recently, the cloning of Dolly the sheep using cells obtained from the
udder of an adult sheep, further undermined this principle in relation to
mammals (Wilmut *et al.*, 1997). For a useful discussion, see Webster and
Goodwin (1996).

10 For example, in the archaeology of domestication (Clutton-Brock, 1989),
histories of colonial exploration (Grove, 1995) and botanical science (Drayton,
2000), and, more recently, the accounts of ethnobotany (Balick and Cox,
1996).

11 As Harlan (1971) points out, the Vavilovian cartography of centres of agricultural origin and diversity that has become such a landmark almost invariably incorporates significant amendments made by his collaborator
Zhukovsky, as in the case of the FAO version reproduced here. His expansion
of the number and extent of these centres, some defining a whole continent,
exhausted the meaningfulness of the 'centres' concept.

12 According to an FAO survey of these *ex-situ* germplasm collections in 1994,
just over half of all accessions were held in genebanks in industrialized
countries, about one-third in developing countries and the remainder in the
international network of IARCs administered by CGIAR (Consultative Group
for International Agricultural Research) (CPGR, 1994b). CGIAR was established in 1971 with funding from the Rockerfeller and Ford Foundations, the
FAO, the World Bank and the UN Development Programme (CPGR, 1987a:
10–15).

13 The global commons are taken to describe those environmental phenomena
and spaces that do not fall under national jurisdiction or private property
rights, notably oceans, Antarctica, the earth's biosphere and outer space, all of
which are the subject of more less established international treaties or protocols (Buck, 1998). They are constituted in international law as a 'common
heritage of [hu]mankind (CHM), a legal concept first established in the UN
Law of the Sea in 1967 (Kotz, 1976) that has four main principles (see
Larschan and Brennan, 1983). The designated commons should:

  (i)   not be subject to appropriation;
  (ii) involve all nations in its management;
  (iii) actively share in any benefits;
  (iv) be dedicated to exclusively peaceful purposes.

14 The 'tragic' rendition of the commons can be traced to the eighteenth/ nineteenth century enclosure of the English commons when 'improvement' and the 'national interest' cast local commons as a hindrance to commercial and state interests (Xenos, 1989). First articulated by propagandists for parliamentary enclosure, like Thomas Malthus (Thompson, 1991: 107), the 'tragedy of the commons' has been deployed more recently in the project of 'third world development' (Roberts and Emel, 1992). The most influential exponent of this twentieth-century variant is Garrett Hardin, for whom 'freedom in a commons brings ruin to all' (1968: 124).

15 The 'virtuous' currency of the commons can be traced back to the natural law tradition of seventeenth-century political debate, where the 'state of nature' signified a gift from God invested in 'man' as an earthly commonwealth (see Squadrito, 1979; Shapiro I., 1991). The dilemma for the theological cast of this early modern political commentary centred on reconciling the nascent figure of the autonomous individual with the moral economy of commonwealth (see, for example, Chalk, 1991). The most influential exposition of this fraught reconciliation is found in the work of John Locke, particularly his *Two treatises on government* (1988/1690), which is still a touchstone of liberal democratic political theory (see Tully, 1980).

16 The 'Agreed Interpretation' (Resolution C 4/89) was formally incorporated as an annex (annex I) to the International Undertaking (Resolution C8/83), securing it as an integral part of its legal provisions.

17 The shift in definitional emphasis from the 'vegetative and propogating' materials of plants to the 'biological diversity of plant genes, etc.' was significant but largely unnoticed in the collective ontology of PGR.

18 The report found that the legal status of materials held in *ex situ* collections is determined primarily on the principles of law and the legislation of the state where the collection is located and, hence, was inevitably varied. The situation for the CGIAR-administered network of International collections was even more complicated with, for example, IARCs exhibiting a wide range of constitutional arrangements. These findings were based on information restricted to those public genebanks that responded to FAO enquiries and excluded private collections (CPGR, 1987a).

19 The grand design of an 'international genebank' envisaged by the FAO's Committee on Agriculture (COAG) as part of an International Convention for PGR back in 1983, complete with floor plans and budget estimates (FAO, 1983c), bore little relation to the arrangements secured by the Undertaking some ten years later. The only significant gesture towards this global jurisdiction was the transfer of the administration of the IARC network of genebanks from CGIAR to the International Board for Plant Genetic Resources (IBPGR) in 1994; itself an FAO satellite organization set up in 1974 (see Kloppenburg, 1988: 161–7).

20 Unsurprisingly the research literature is no less confused than the FAO Conference records on the applicability of the 'common heritage of humankind' principle to the International Undertaking (e.g. Juma, 1989; Ramakrishna, 1992; Flitner, 1998).

21 This distinction between physical (or tangible) and intangible property in western jurisprudence is indebted to that made in Roman law between corporeal and incorporeal things (Drahos, 1996).

22 The criteria of 'invention' exercised in patent law, for example, require that the knowledge-object is useful (i.e. it must have an industrial or commercial application); novel (i.e. it must be original and not already known in the public domain); and non-obvious (i.e. it must be more inventive than 'mere discovery' of something that already exists) (Cornish, 1999). The US Supreme Court ruling in *Diamond* v. *Chakrabarty* (1980) is now taken as the landmark decision effecting the ontological shift that admitted microbiological knowledge practices and knowledge-objects as patentable 'inventions'.

23 At the time of the International Undertaking on PGR no developing countries had implemented legislation along UPOV lines and none was a member of the Union. This remained the case until Latin American countries party to establishing the North American Free Trade Area were obliged to join UPOV as a condition of NAFTA membership. Under the original terms of the UPOV Convention, farmers re-using seeds from their own harvest and plant breeders' using protected varieties to produce further improvements were exempt from the prohibition on third-party use of protected varieties. These exemptions were revoked in the 1991 revisions to bring UPOV into line with Trade Related aspects of Intellectual Property Rights (TRIPS) provisions being constituted by the World Trade Organization (see Correa, 1995) (see note 24 below).

24 This shift was extended to plant varieties by the European Patent Office in 1983 in its reinterpretation of Article 53 of the European Patent Convention (1973) which had specifically excluded 'plant or animal varieties or essentially biological processes for the production of plants or animals' from the scope of permissible patent claims. In a ruling on a case brought to the appeal board by Ciba-Geigy, varieties produced by genetic engineering were deemed to be 'novel plants' rather than 'new varieties' and hence patentable. This principle has since been incorporated into revisions to the UPOV Convention (1991). For overviews of these developments in IPR in relation to living things contemporaneous with the event under discussion see Sedjo, 1992 (for a US perspective) and Bergmans, 1991 (for a European perspective).

25 Under the Farmers' rights provisions of the Undertaking, entitlements to compensation were vested in the FAO as a 'trustee of the international community' to fund community seedbanks and traditional land management and conservation practices. In practice, its trusteeship suffered from under-funding, as financial contributions to the International Fund failed to materialize, and was contested by some member countries who wanted it to be vested in nation states (e.g. Turkey; see FAO, 1989b). It is also worth reiterating that the purchase of Farmers' rights as 'rights' was compromised by the non-binding or 'soft-law' status of the Undertaking (Correa, 1995).

26 The itch of the primitive that consigned indigenous peoples to 'the state of nature', expelling them from the compass of the social in imperial mappings of *terra nullius*, persists today in their incorporation into the genomic mappings of the life sciences as subjects of bio-prospecting (Cunningham, 1998). The extension of IPR to the patenting of such genomic 'inventions' is itself intensely contested within scientific and legal communities (see Gannon *et al.*, 1995; Black, 1998)

27 Lipietz describes the Rio Summit as 'a diplomatic Vietnam' for the Bush (senior) administration (1995: 9). The Clinton administration signed the

Convention on Biological Diversity in summer 1993, but Congress refused to ratify it in summer 1994. It remains unsigned by the USA.

Section 3: Introduction

1 A similar point is made by Derrida in a conversation with J.L. Nancy where he explains his rare treatment of 'the subject' as a product of the fact that 'the discourse of the subject, even if it locates difference . . . continues to link subjectivity with man' (1991: 105). Moreover, Appadurai himself makes the limits of his project in *The social life of things* (1986) clear in an interview in which he describes it as an effort 'to milk the conceit that we need to forget people for a moment and think of things themselves, as in some kind of way having a *life*' (Bell, 2000: 27, original emphasis).

chapter 6

1 This is not to suggest that food-born toxicities and diseases are peculiar to the industrial era. Adulterated and rotten foodstuffs were historically common-place, and industrial preservation and transport technologies have extended their compass and durability (see Kiple and Ornelas, 2000).

2 It is worth noting how frequently scientific authorities evoke the term 'rogue' to explain food scares to the wider public through the media. For example, in the case of BSE–vCJD, the prions implicated in transmissable spongiform encephalopathies (TSEs) were characterized as 'rogue proteins', and in the food poisoning outbreak in Lanarkshire in 1997, the bacterial strain *E. coli* 0157 was identified as the 'rogue bacteria' responsible. It seems that such potent agents only emerge into the glare of public acclaim at moments of rupture in the disciplinary practices and accounts of science.

3 Michel Serres makes much of the hyphen's replacement of the inverted capitalized omega as the conventional sign of union or connection between two words. He argues that it imprints diacritically the meaning of the 'middle ground' or 'excluded third', thus acting as a shorthand for his various metaphorical figures of absent presence – the parasite; the tiers instruit; the blank figure of the joker (see Assad, 1999: 132–4). Deleuze takes a similar tack when he insists on the significance of the conjunction 'and' in his mode of thinking, which he argues 'has enough force to shake and uproot the verb "to be" ' emphasizing between-ness as 'another way of travelling, . . . coming and going rather than starting and finishing' (Deleuze and Guattari, 1988: 25).

4 Exceptions which stretch across economic/cultural, production/consumption divides are growing in number in response to the practical imperatives of addressing the crises in farm livelihoods and consumer confidence engendered by industrialization. See, for example, Whatmore and Thorne (1997) on Fairtrade networks; Cook and Crang (1998a, 1998b) on consumer under-standings of the origins of foods; Fitzsimmons and Goodman (1998) on alternative food networks; and Murdoch *et al.* (2000) on the construction of 'quality' foods.

5 Exceptions to this exclusive focus on the meaningfulness of foods in (human) cultural practices include 'popular' bio-graphies of staple foodstuffs and their

active role in human history, e.g. Zuckerman (1998) on the potato; and Hobhouse (1999) on tea and sugar.

6 An exception to this is Adams's gestures towards a 'feminist-vegetarian agenda' (1990).

7 As Deleuze and Guattari note (1988: 519, footnote 13), this contrast between arborescent and rhizomatic forms of thinking is also taken up by Michel Serres (1975) to rather different effect in his examination of the networking assemblage of the tree itself and its use in a variety of scientific domains.

8 The Soya Bean Information Centre (whose publicity does not acknowledge that it is funded by Monsanto) claims in an information sheet that 'experts estimate that up to 30,000 food products contain soya derivatives as ingredients' (Monsanto, 1996). As well as a host of human and non-human foodstuffs, Hapgood illustrates the soybean's versatility in his *National Geographic* article (1987) with a painting by the artist James Gurney which includes more than 60 soy products – from glues and petrol, fire hydrant foam and plastic, to the paint used in the painting itself.

9 For parallels with Merleau-Ponty's visible/invisible and his concerns with the presentation of absence, see Carey (2000: 31–2).

10 The Monsanto Corporation is not unfamiliar with controversy in relation to its products. These include the defoliant Agent Orange, used extensively in the Vietnam War, and the bovine growth hormone rBST (recombinant bovine somatotropin) designed to boost milk yields and banned by the European Union (Palast, 1999).

11 The high protein and fatty acid components of soya make it a valuable component of vegetarian alternatives to meat and dairy products in today's industrial diet. However, health claims made for these products have been called into question by scientists breaking ranks with the US Food and Drug Administration's position. They point to the toxicity of the oestrogen-like properties of isoflavones in the bio-chemistry of soya (Fallon and Enig, 2000; Institute of Food Research, 2000).

12 The modern soybean taxon *Glycine max* is located as a sub-species of the tribe (*Phaseolae*) within the largest of the three *Leguminosae* sub-families (*Papilionoideae*) (ILDIS, 2000).

13 North-east China was recognized by the early twentieth-century Soviet botanist Vavilov as one of the eight 'centres of origin' for the world's most important crop plants, which have since become closely allied to the identification of 'centres of plant genetic diversity' (see chapter 5).

14 These nitrogen-fixing root nodules are general in two of the three sub-families of *Leguminosae* (*Mimosoideae* and *Papilionoideae*) but rare in the third (*Caesalpinioideae*) (ILDIS, 2000).

15 To this extent the soybean, and leguminous plants more generally, are excellent illustrations of new arguments in evolutionary theory which cast the genealogical emphasis of Darwinian ideas in a much more relational light, either in terms of the 'co-evolution' of different species (see Eldridge, 1995) or, more radically, of evolutionary symbiosis at the cellular level (see Margulis and Fester, 1991).

16 The soybean's earlier presence in Europe was recorded in 1737 by the Swedish botanist Carolus Linnaeus in an inventory of plants growing in a garden in Holland (see Hapgood, 1987: 79). Another US Naval officer, Matthew Perry, is

credited with its introduction to the west in 1825 (Kiple and Ornelas, 2000: 1855).

17 Until the advent of GM varieties, nearly all of today's US soybean crop can be traced as descendants of just six plants from this period of concerted acquisition (Fowler and Mooney, 1990: 84).

18 Kloppenberg gives the example of seed-corn firms employing some 125,000 labourers over a 2–4 week period to de-tassle their breeding crop (1988: 112).

19 Unlike animals, the cellular potential of plants is highly plastic with a large numbers of cells that are *totipotent*, that is that have the potential to generate an entire plant with all its various tissues (Tudge, 1993: 186–7).

20 The United States, China, Brazil and Argentina account for over 90 per cent of the global soybean crop today.

21 Glyphosate is the active ingredient of Monsanto's Roundup® herbicide, typically comprising 41 per cent. Glyphosate (N-phosphonomethyl glycine) is a post-emergence broad spectrum herbicide that kills all green plants not engineered to tolerate it. It works by inhibiting aromatic amino acid biosynthesis in the leaf chloroplasts, specifically the enzyme 5-enol-pyruvylshikimicacid 3-phosphate (EPSP) synthase, thereby disabling the conversion of light into chemical energy and so preventing plant growth (Schulz *et al.*, 1990: 7–8). The other 59 per cent of Roundup® is made up of a range of 'inert' ingredients, including Polyethyloxylated tallow amine surfactant (POEA) to de-clog applicators and facilitate even spray coverage, which harbour toxicities of their own (Lappé and Bailey, 1999: 54). Monsanto's monopoly on commercial glyphosate herbicide through its Roundup® brand patent ran out in 2000. The corporation is already involved in legal suits in the United States to prevent its Swiss competitor AstraZeneca from testing their rival glyphosate herbicide brand Touchdown® on its Roundup Ready™ soybeans (*Daily Telegraph*, 22/1/99).

22 It is worth noting that Comai's research team was employed by Calgene Inc., one of the US pioneers in biotechnology (acquired by Monsanto in 1997). The prevalence of corporate scientists among the authors of research articles in key journals like *Bio/Technology*, *Trends in Biotechnology*, *Science* and *Nature* is remarkable (at least to this outsider), but is rarely the subject of reflection, let alone concern, within these pages. This commercialization of biotechnological science has effectively obscured the distinction between 'pure' and 'applied' research and compromised the self-proclaimed 'objectivity' of scientists as their practices and results 'disappear' behind the cordon of commercial confidentiality (Rothman *et al.*, 1996).

23 In addition to Roundup Ready™ soybeans, Monsanto have also patented GM tobacco, cotton, sugar beet, corn (maize), and canola (oilseed rape) seed. Ironically, Agracetus Inc. was awarded a patent granting rights to all forms of genetically engineered soybeans by the European Patent Office in March 1994, a patent that has been hotly challenged by Monsanto among other agri-biotechnology corporations (Stone, 1995).

24 Monsanto is actively (and successfully) prosecuting farmers, mainly in the US mid-west and Canadian prairies, whom it suspects of acquiring its herbicide or GMHT seed outside the terms of this contract. Pinkertons private investigation agency has been employed to secure evidence on its behalf (*Washington Post*,

3/2/99 and 2/5/99; *Independent on Sunday*, 14/3/99). The logical trajectory of this monopoly impulse is harboured in the infamous 'terminator technology' owned by Monsanto which could be used to 'switch off' the germinal properties of Roundup Ready® seeds and effectively sterilize them to prevent 'unauthorized' use.

25 The anti-GM scientist Mae-Wan Ho (1999: 61) summarizes the theories of biological complexity which counter the reductivist logic of genetic determinism, and their practical consequence for genetic engineering, in terms of three counter-propositions. A gene does not determine a function, rather genes perform in a complex network in which their relationship to characteristics is non-linear and multidimensional. Genes and genomes are not stable and unchanging, rather they are dynamic and fluid, generating 'adaptive' mutations in particular environmental contexts. Genes do not stay where they are put, rather they 'move' within and between species, recombining in unintended ways. These propositions are now finding their way from the critical 'fringes' of the scientific community to the emerging orthodoxies of post-genomic research (see Sarkar, 1998).

26 Like other organophosphate pesticides, glyphosate's rate of environmental decomposition varies dramatically from the experimental *in vitro* conditions of the laboratory and the field test site, to the variable *in vivo* conditions of particular soils, hydrological environments and farming practices. Moreover, they have proved to be toxic to the nervous and immune systems of mammals (Raganarsdottir, 2000). Monsanto's successful applications in the USA, UK and Australia to triple the permissible level of glyphosate residues in its Roundup Ready™ soybeans from six to 20 parts per million (*Nature Biotechnology* (1997), 15: 1233) suggests that claimed reductions in volume of application will be tempered by increased intensity of application. The UK's Pesticide Safety Directorate report (1999) on GMHT crops concluded that 'there is currently a lack of independent research to allow an accurate prediction of the potential impacts' of GMHT technologies on pesticide use and its environmental consequences (executive summary, point 5).

27 Prior approval for the commercial or large-scale release of GMOs was first required in 1997. At this time, the Advisory Committee on Releases to the Environment (ACRE) was responsible for licensing experimental planting in the UK, while the Advisory Committee on Pesticides (ACP) was responsible for advising on the safety of pesticide products, usage and residues. English Nature, the statutory body responsible for nature conservation in England led the science/policy opposition to the failure of these advisory committees to address wildlife or biodiversity concerns within their remits. This opposition arose, at least in part, in response to Monsanto's efforts to secure a licence for field trials in Roundup Ready™ oil-seed rape (English Nature, 1998).

28 The main UK precedent for the licensing of a derivative transgenic food product at the time was the treatment of a tomato paste made from the GM Flavrsavr™ variety. This product was licensed by the Advisory Committee on Novel Foods and Processes (ACNFP) in 1997 and labelling was required by the Food Advisory Committee (FAC). It entered the UK marketplace clearly labelled as a GM product and caused little public outcry or consumer backlash as a result. Since then, the patent rights on Flavrsavr™ tomatoes have passed to Monsanto with its takeover of rival biotech company Calgene in 1997.

29 These advertisements were produced by the London-based agency Bartle, Bogle and Hegarty. They were subsequently condemned by the British Advertising Standards Agency for representing Monsanto's environmental claims about its GMHT products in 'confusing', 'misleading', 'unproven' and 'wrong' ways, and expressing the corporation's own opinion of transgenic engineering as an extension of traditional plant breeding methods as 'accepted fact', when it is a matter of dispute within the scientific community (*Guardian*, 1/3/99; *Living Earth*, 1999, 202: 8).

30 The Advisory Committee on Novel Foods and Processes relied on animal trials of Roundup Ready® soybeans lasting no more than a matter of months to satisfy their 'test' requirements for impacts on human health (Derek Burke, Chair of ACNFP, interviewed on BBC television's *Panorama* programme, 18/5/1999). The adequacy of animal analogues for 'testing' the toxicological effects of human foods is an acknowledged problem among OECD government scientists (New Scientist, 1999).

31 It is noteworthy that a subsequent study by the Advisory Committee on Animal Feeding Stuffs, which reports to the Food Standards Agency, found that 'DNA fragments large enough to contain potentially functional genes survived processing in many of the samples [of animal feedstuffs] studied' (*Observer*, 15/10/2000). Animal feeds are the primary destination of soya meal and UK retailers are currently extending their GM soya ban to their meat supply networks while the Food Standards Agency is pressing for the compulsory labelling of food derivatives from animals raised on GM feed.

32 It is a marketing strategy that Monsanto had used before with its rBHT (Bovine Growth Hormone), patented under the brand name Posilac®, where it fought against European (and US) efforts to separate rBHT milk and conventional milk and label products derived from hormone-treated milk. The corporation's rationale for its dogged and politically damaging resistance to the segregation and labelling of GMHT soybeans was economic. Labelling would 'stigmatize' the product and its derivatives, associating it with 'risk' in the public mind and segregation would permit this perception to find expression through market choice.

33 These included a Mori Poll in 1996 (on behalf of Greenpeace) and a widely leaked report by Stanley Greenberg for Monsanto that showed public opposition to GM crops rising from 38–50 per cent between 1997 and 1998 in the UK, and from 47–57 per cent among AB social classes (Ford, 2000: 77).

34 Unlike quality assurance systems, 'product traceability' is more closely allied to the certification practices of alternative food networks like organics or Fairtrade.

35 Major fast-food chains and local education authorities responsible for school meals joined food retailers and processors in banning GM ingredients. For a telling exposition of the US industry's position on labelling, see Miller H. (1999).

36 These include genetic 'fingerprinting' techniques that can detect modifications in soya products to a fraction of 1 per cent and the extension of current segregation practices (e.g. for beans of different protein or oil content) to non-GM soya with a minimal price premium (see Buckwell and Brookes, 1999).

37 It is important not to exaggerate this political gesture in view of the fact that 11 of the 13 members of ACRE were not eligible to renew their committee membership anyway.

38 One is reminded here of Rachel Carson's careful, passionate science and the way in which her opposition to the programmatic use of the pesticide DDT earned her personal and professional vilification by corporate and government scientists who questioned her scientific credentials as an 'unmarried woman' and 'probable communist' (Lear, 1997: 428–35).

chapter 7

1 Notable reworkings of the natural law tradition include those of Aquinas and Grotius, but Locke's work best epitomizes early modern tensions between notions of 'common good' and 'individual good' (see Tuck, 1979; Tully, 1993).

2 Contemporary writers in this Kantian tradition have modified their reliance on the impartiality of justice by recognizing that competent moral agents are contracted on unequal terms; a theme pursued most influentially by John Rawls (1971) in his 'difference principle', and by Will Kymlicka (1991) in his notion of the 'pluralist contract'.

3 Persons in law can be non-individuals, for example states, corporations, unions etc. McHugh has argued that if the concept of the 'security of the individual' (central to human rights law) were extended from persons to human beings, this would contribute towards the realization of substantive equality (i.e. in terms of the material prerequisites for participating as equal members of a polity) (1992: 460).

4 It is no coincidence that the language of early women's struggles for political rights, notably in the writings of Mary Wollstonecraft, should borrow from those for the abolition of slavery in likening the status of wives to that of slaves (see Ferguson, 1992).

5 This is not to suggest that these are the only responses (for example, Habermasian critical theory is also notable) but rather that they have been the most influential in the sense of being translated into discourses beyond the academy.

6 Interestingly, Mouffe points to similar problems to those raised here with what she calls 'a certain type of extreme postmodern fragmentation of the social' (1995: 262) – but without identifying any alleged 'extremists'.

7 See also Diprose's notion of 'corporeal schema' which takes up Merleau-Ponty's idea of the body's directional activity or 'intentional arc' (1994: 106) and the special issue of *Hypatia* on 'feminism and the body' (fall, 1991).

8 See also, Leder (1990a, 1990b) and Levin (1990).

9 The ethical standing of animals has been a matter of longstanding dispute in moral philosophy, well in advance of contemporary environmentalism. Particularly influential contributions include the Thomist legacy of Thomas Aquinas in the natural law tradition and the utilitarian legacy of Jeremy Bentham in the social contract tradition (see Midgley, 1983). For an excellent fictional rendition of these philosophical arguments see Coetzee's *The lives of animals* (1999).

10 A good example is the global network DAWN (Development with Women working for a New Era) which since 1984 has sought to articulate material connectivities between environmental, livelihood and health issues and the centrality of 'third world' women in this nexus (Braidotti *et al.*, 1994).

11   Ansell-Pearson (1997) provides a useful account of Deleuze and Guattari's acknowledged debt to Bergson's philosophical account of creative evolution (1983/1907) and the biologist von Uexküll's contrapuntal conception of biological processes and forms (1992/1934) (see also Ingold, 1995a).

12   Obviously this is not to suggest that Latour is not passionate about his work. One has only to think of the title and style of his book about the unrealized blueprint for a rapid transport system in Paris – *Aramis or the love of technology* (1996), or his zealous efforts to ally science and science studies (1999a) against their caricatured enmity in the so-called 'science wars'. But rather that his work is not passionate in the sense taken in this book from Game and Metcalf's *Passionate sociology*, namely that he 'masterfully refuse[s] to place [himself] within the social life [he] studies' (1996: 5). In so far as he positions himself beyond the academy at all it is by dissociation. For example, his aversion to 'a conception of left-wing radicalism that has not yet been renewed as forcefully as science has been' (1997b: xvii); or his repudiation of the misguided terms on which 'green' parties and movements have sought to put 'nature' on the political agenda (1999c).

13   Ingold's 'weaving'/'making' variant of the Heideggerian distinction between 'dwelling' and 'building' purposefully rejects its insistence that human rationality and subjectivity mark an absolute break from the animal world (see also Glendinning, 1998: 73–4).

14   These transpecies infectivities were not limited to cattle and humans but have been recorded in increasing numbers in companion animals (notably cats) and zoo animals (notably deer), giving rise to the generic term TSEs (transpecies spongiform encephalopathies) (Ridley and Baker, 1998).

15   The most exhaustive account of the shifting sands of Government policy and scientific advice towards BSE–vCJD in the 1980s–1990s and assessment of the distribution of responsibility for its devastating failings is provided by the voluminous report of the Lord Phillips' enquiry into BSE (see BSE Inquiry, 1999 and associated website).

16   'Couplings', like 'cyborgs', betoken a version of hybridity in which difference is prefigured in the alterity of already constituted kinds. By contrast, the emphasis in my account on the indeterminacy of difference draws on Bergson's bio-philosophy, particularly his notion of *differentiation* as an explosive 'internal' life force (1983/1907), subsequently taken up and reworked by Deleuze (1994/1968) (and with Guattari (1988/1980)). This distinction is important in understanding the contrast between, say, the approaches of Latour and Haraway to hybridity. For valuable discussion on these points, see Ansell-Pearson (1999: 33–69) and Hansen (2000b).

References

Aboriginal Law Bulletin, 1993. Speech of the Honourable Prime Minister, Paul Keating MP. Australian Launch of the International Year for the World's Indigenous People. *Aboriginal Law Bulletin*, 3/61: 4–5.

Abram D., 1988. Merleau-Ponty and the voice of the earth. *Environmental Ethics*, fall: 110–25.

Abram D., 1997. *The spell of the sensuous*. Pantheon Books, New York.

Ackerman B., 1977. *Private property and the constitution*. Yale University Press, New Haven, CT.

Adams C., 1990. *The sexual politics of meat*. Polity Press, Oxford.

Allaire G. and R. Boyer (eds), 1995. *La grand transformation*. INRA, Paris.

(ALR) Australian Law Reports, 1992. Mabo v. Queensland. *Australian Law Journal*, 66: 408–99.

American Soybean Association, 1996. *European response to GM soybeans*. *www.oilseeds.org/asa/news.htm*

Anderson A., 1992. Cryptonormativism and double gestures: the politics of poststructuralism. *Cultural Critique*, Spring: 63–95.

Anderson D., 1998. Property as a way of knowing on Evenki land in Arctic Siberia. In Hann C. (ed.), *Property relations*: 64–84. Cambridge University Press, Cambridge.

Anderson K., 1997. A walk on the wildside: a critical geography of domestication. *Progress in Human Geography*, 21/4: 463–85.

Anderson K., 1999. Science and the savage: the Linnean Society of New South Wales, 1874–1900. *Ecumene*, 6: 125–43.

Anderson K., 2000. 'The beast within': race, humanity, and animality. *Society and Space*, 18/3: 01–320.

Anderson L., 1999. *Genetic engineering, food and our environment*. Green Books, Totnes.

Ansell-Pearson K., 1997. *Viroid life. Perspectives on Nietzsche and the transhuman condition*. Routledge, London.

Ansell-Pearson K., 1999. *Germinal life. The difference and repetition of Deleuze*. Routledge, London.

Appadurai A. (ed.), 1986. *The social life of things. Commodities in cultural perspective*. Cambridge University Press, Cambridge.

Appadurai A., 1996. *Modernity at large. Cultural dimensions of globalization*. Minnesota University Press, Minneapolis, MN.

Arluke A. and C. Sanders, 1996. *Regarding animals*. Temple University Press, Philadelphia, PA.

Arnell B., 1996. *John Locke and America. The defence of English colonialism*. Clarendon Press, Oxford.

Assad M., 1999. *Reading with Michel Serres. An encounter with time*. SUNY Press, Albany, NY.

Attwood B. and A. Markus, 1999. Representation matters: the 1967 referendum and citizenship. In Peterson N. and W. Sanders (eds), *Citizenship and indigenous Australians*: 118–40. Cambridge University Press, Cambridge.

Aubertin C. and F.-D. Vivien, 1998. *Les enjeux de la biodiversité*. Economica, Paris.

Auguet R., 1972. *Cruelty and civilization. The Roman games*. Routledge, London.

Austin A., 1999. (Member of Advisory Commission on Novel Foods and Processes) speaking on *Panorama* documentary on the GM food controversy, 18/5/99. BBC, London.

Balick M. and P. Cox, 1996. *Plants, people and culture. The science of ethnobotany*. Scientific American Library, New York.

Bao X. *et al.*, 1993. *Glycine max* and wild soybean relatives. *Plant Genetic Resources Newsletter* 94: 1–3. FAO/IBPGR, Rome.

Barad K., 1998. Getting real: performativity, materiality and technoscientific practices. *Differences*, 10/2.

Barad K., 1999. Agential realism. Feminist interventions in understanding scientific practices. In Biagioli M. (ed.), *The Science Studies Reader*: 1–11. Routledge, London.

Barnes, T. and J. Duncan (eds), 1992. *Writing worlds. Discourse, text and metaphor in the representation of landscape*. Routledge, London.

Barrère M. (ed.), 1992. *Terre patrimoine commun*. La Découverte/Association Descartes, Paris.

Bartlett R., 1993. The source, content and proof of Native Title at common law. In Bartlett R. (ed.), *Resource development and aboriginal land rights in Australia*: 35–60. The Centre for Commercial and Resources Law, The University of Western Australia and Murdoch University, Perth, WA.

Bateson G., 2000 (1st edition 1972). *Steps to an ecology of mind*. Chicago University Press, Chicago, IL.

Bateson P., 1997. *The behavioural and physiological effects of culling red deer*. National Trust, London.

Battaglia D., 1994. Retaining reality: some practical problems with objects as property. *Man*, 29/4: 631–44.

Bauman Z., 1992. *Mortality, immortality and other life strategies*. Polity Press, Cambridge.

Bauman Z., 1996. *Life in fragments*. Polity Press, Cambridge.

Beck U., 1989. *The risk society*. Sage, London.

Bell D. and G. Valentine, 1997. *Consuming geographies*. Routledge, London.

Bell V., 2000. Historical memory, global movements and violence: Paul Gilroy and Arjun Appadurai in conversation. In Bell V. (ed.), *Performativity and belonging*: 21–40. Sage, London.

Benhabib S., 1987. General and concrete others. In Benhabib S. and D. Cornell (eds), *Feminism as critique*: 77–96. Polity Press, Cambridge.

Benton T., 1993. *Natural relations. Ecology, animal rights and social justice*. Verso, London.

Benton T. (ed.), 1996. *The greening of Marxism.* Guilford Press, New York.

Berg M. and A. Mol, 1998. *Differences in medicine: Unravelling practices, techniques and bodies.* Duke University Press, Durham, NC.

Bergmans B., 1991. *La protection des innnovations biologiques. Une étude de droit comparé.* Larcier, Brussels.

Bergson H., 1983 (1907 in French). *Creative evolution* (English trans. A. Mitchell). University Press of America, Lanham, MD.

Berlan J.-P. and R. Lewontin, 1986. Breeders rights and patenting life forms. *Nature,* 322/August: 785–9.

Béteille A., 1998. The idea of indigenous people. *Current Anthropology,* 39: 187–91.

Bhabha H., 1994. *The location of culture.* Routledge, London.

Bhabha H., 1997. The world and the home. In McClintock A., A. Mufti and E. Shohat (eds), *Dangerous liaisons. Gender, nation and postcolonial perspectives:* 445–55. Minnesota University Press, Minneapolis, MN.

Bigwood C., 1993. *Earth Muse: feminism, nature and art.* Temple University Press, Philadelphia, PA.

Bingham N., 1996. Object-ions: from technological determinism towards geographies of relations. *Society and Space,* 14: 635–57.

Bingham N., 2001. In the belly of the monster: Frankenstein, food, factishes, and fiction. In Kitchen R. and J. Kneale (eds), *The spaces of science-fiction.* Routledge, London.

Bingham N. and N. Thrift, 2000. Some new instructions for travellers: the geography of Bruno Latour and Michel Serres. In Crang M. and N. Thrift (eds), *Thinking space:* 281–301. Routledge, London.

Birch T., 1990. The incarceration of wildness: wilderness areas as prisons. *Environmental Ethics,* 12: 3–26.

Bird E., 1987. The social construction of nature: theoretical approaches to the history of environmental problems. *Environmental Review,* 11: 255–64.

Black J., 1998. Regulation as facilitation: negotiating the genetic revolution. *Modern Law Review,* 61/5: 49–88.

Blackstone W. Sir, 1803 (1st published 1765). *Commentaries on the laws of England* (2 volumes). Strachan, London.

Blomley N., 1994. *Law, space and the geographies of power.* The Guilford Press, New York.

Blomley N., D. Delaney and R. Ford (eds), 2001. *The legal geographies reader.* Basil Blackwell, Oxford.

Bonner J., 1974. *On development: the biology of form.* Harvard University Press, Cambridge, MA.

Bora A., 1999. Discourse formations and constellations of conflict: participation in the German debate on genetically altered plants. In O'Mahony P. (ed.), *Nature, risk and responsibility. Discourses of biotechnology:* 130–46. Macmillan, Basingstoke.

Boulter D., 1997. Scientific and public perception of plant genetic manipulation – a critical review. *Critical Reviews in Plant Sciences,* 16/3: 231–51.

Bowker G. and S. Leigh Star, 1999. *Sorting things out. Classification and its consequences.* MIT Press, Cambridge, MA.

(BPP) British Parliamentary Papers, 1836–37. *Report and minutes of evidence of the House of Commons Select Committee on Aborigines (British Settlements).*

Braidotti R., *et al.*, 1994. *Women, the environment and sustainable development. Towards a theoretical synthesis.* Zed Books, London.

Braun B. and N. Castree (eds), 1998. *Remaking reality. Nature at the millennium.* Routledge, London.

Brennan F., 1991. *Sharing the country.* Penguin Australia, Ringwood, Victoria.

Bromley D. (ed.), 1993. *Making the commons work.* Institute of Contemporary Studies, San Francisco, CA.

Brown S. and R. Capdevilla, 1999. *Perpetuum mobile*: substance, force and the sociology of translation. In Law J. and J. Hassard (eds), *ANT and after*: 26–50. Basil Blackwell, Oxford.

BSE Inquiry, 1999. *The BSE Inquiry* (16 volumes). HM Stationary Office, London. See also www.bse.org.uk

Buchanan I., 1997. The question of the body in Deleuze and Guattari: or, what can a body do? *Body & Society*, 3/3: 73–91.

Buck S., 1998. *The global commons. An introduction.* Earthscan, London.

Buckle S., 1991. Natural law. In Singer P. (ed.), *A companion to ethics*: 161–74. Basil Blackwell, Oxford.

Buckwell A. and G. Brookes, 1999. *The cost of segregation of GM and non-GM crops.* Wye College, University of London.

Buell L., 1995. *The environmental imagination.* Harvard University Press, Cambridge, MA.

Burke D., 1998. The 'yuk' factor. In Griffiths S. and J. Wallace (eds), *Consuming passions*: 48–57. Mandolin, Manchester.

Busch L., W. Lacy, J. Burkhardt and L. Lacy, 1991. *Plants, power and profit. Social, economic and ethical consequences of the new biotechnologies.* Basil Blackwell, Oxford.

Callicott J. Baird, 1979. Elements of an environmental ethic: moral considerablity and the biotic community. *Environmental Ethics*, 1: 71–81.

Callicott J. Baird, 1989. *In defence of the land ethic: essays on environmental philosophy.* State University of New York Press, Albany, NY.

Callon M., 1986. Some elements of a sociology of translation: domestication of the scallops and fishermen of St Brieuc Bay. In J. Law (ed.), *Power, action, belief: a new sociology of knowledge?* RKP, London.

Callon M., 1992. Techno-economic networks and irreversibility. In Law J. (ed.), *A sociology of monsters*: 196–229. Routledge, London.

Callon M. (ed.), 1998. Introduction to *The laws of markets*: 1–57. Basil Blackwell, Oxford.

Callon M. and B. Latour, 1981. Unscrewing the big leviathan. In Knorr-Cetina K. and A. Cicourel (eds), *Advances in social theory and methodology*: 83–103. RKP, London.

Callon M. and J. Law, 1995. Agency and the hybrid collectif. *South Atlantic Quarterly*, 94/2: 481–507.

Canguilhem M., 1996. Machine and organism. In Kwinter J. and P. Crary (eds), *Incorporations*: 45–68. Zone Books, New York.

Caraway N., 1991. The cunning of history: empire, identity and feminist theory in the flesh. *Women and Politics*, 12/2: 1–18.

Carey S., 2000. Cultivating ethos through the body. *Human Studies*, 23: 23–42.

Carter P., 1987. *The road to Botany Bay.* Faber and Faber, London.

Casey E., 1998. *The fate of place. A philosophical history.* University of California Press, Berkeley, CA.

Castree N., 1996. Birds, mice and geography: Marxisms and dialectics. *Transactions of the Institute of British Geographers*, 21/2: 342–62.

Cavalieri P. and P. Singer (eds), 1993. *The great ape project: equality beyond humanity*. The Fourth Estate, London.

(CBD), *Convention on Biological Diversity*, 1992. United Nations Environment Programme, Nairobi.

de Certeau M., 1988 (1984). *The practice of everyday life* (trans. S. Rendall). University of California Press, Berkeley, CA.

de Certeau M., 1985. Montaigne's 'Of Cannibals': the savage 'I'. In *Heterologies: discourse on the other* (trans. B. Massumi). University of Minnesota Press, Minneapolis, MN.

de Certeau M., L. Giard and P. Mayol, 1998. *The practice of everyday life. Volume 2: Living and Cooking* (trans. T. Tomasik). University of Minnesota, Minneapolis, MN.

Chalk A., 1991 (1911). Natural Law and the rise of economic individualism in England. Reprinted in Blaug M. (ed.), *Pre-classical economists*, Vol. 1. Edward Egar, Aldershot.

Chaloupka W. and R. McGreggor Cawley, 1993. The great wild hope. In Bennett J. and W. Chaloupka (eds), *In the nature of things. Language, politics and the environment*: 3–24. Minnesota University Press, Minneapolis, MN.

Charlesworth M., 1924. *Trade routes and commerce in the Roman empire*. Cambridge University Press, Cambridge.

Chatwin B., 1987. *The songlines*. Jonathan Cape, London.

Cheney J., 1989. Postmodern environmental ethics: ethics as bioregional narrative. *Environmental Ethics*, 11/2: 117–34.

(CITES) Convention on International Trade in Endangered Species of Wild Fauna and Flora, 1997a. Document 10.22, *Cooperation/synergy with other conservation conventions and agencies*. 10th meeting of the Parties, 9–20 June, Harare.

(CITES) Convention on International Trade in Endangered Species of Wild Fauna and Flora, 1997b. *Summary report 2nd session Committee Meeting*, 11 June; section XIV, 31. 10th meeting of the Parties, 9–20 June, Harare.

Clark N., 1997. Panic ecology. Nature in the age of superconductivity. *Theory, Culture and Society*, 14/1: 77–96.

Clay J., 2001. *Maconochie's experiment*. John Murray Publishers, Edinburgh.

(CLR) Commonwealth Law Reports, 1996. Wik v. Queensland. *Commonwealth Law Reports*, 187.

Clutton-Brock J. (ed.), 1989. *The walking larder, patterns of domestication, pastoralism and predation*. Unwin & Hyman, London.

Coetzee J., 1999. *The lives of animals*. Princeton University Press, Princeton, NJ.

Cole S., 1997. Do androids pulverise tiger bones to use as aphrodisiacs? In Taylor P, S. Halfron and P. Edwards (eds), *Changing life. Genomes, ecologies, bodies, commodities*: 175–95. Minnesota University Press, Minneapolis, MN.

Coles R., 1993. Eco-tones and environmental ethics. In Bennett J. and W. Chaloupka (eds), *In the nature of things*: 226–49. Minnesota University Press, Minneapolis, MN.

Comai L., D. Facciotti, W. Hiatt, G. Thompson, R. Rose and D. Stalker, 1985a. Expression in plants of a mutant *aroA* gene from *Salmonella typhimurium* confers tolerance to glyphosate. *Nature*, 317: 741–4.

Comai L., D. Facciotti, W. Hiatt, G. Thompson, R. Rose and D. Stalker, 1985b. Expression in plants of a bacterial gene coding for glyphosate resistance. In

Zaitlin M., P. Day and A. Hollaender (eds), *Biotechnology in plant science. Relevance to agriculture in the eighties*: 329–37. Academic Press, Orlando, FL.

Competition Commission, 2000. *Report of enquiry into UK supermarkets*. Department of Trade and Industry. www.competition-commission.gov.uk/446.htm

Cone R. and E. Martin, 1997. Corporeal flows. The immune system, global economies of food and implications for health. *The Ecologist*, 27/3: 107–11.

Conley V., 1997. *Ecopolitics. The environment in post-structuralist thought*. Routledge, London.

Consumers Association, 1999. Written evidence to the House of Lords European Select Committee on European Communities. *Second report on the regulation of genetic modification in agriculture*. www.publications.parliament.uk/pa/ld199899/ldselect/ldeucom/11/11we13

Cook I. and P. Crang, 1998a. The world on a plate. *Journal of Material Culture*, 1/2: 131–53.

Cook I. and P. Crang, 1998b. Biography and geography. Consumer understandings of the origins of foods. *British Food Journal*, 100/3: 162–7.

Cooper D., 1991. Genes for sustainable development. In World Rainforest Movement, *Biodiversity: social and ecological perspectives*: 105–23. Zed Books, London.

Cooper D., 1993. The international undertaking on Plant Genetic Resources. *Review of European Community and International Environmental Law*, 2/2: 158–66.

Cornell D., 1985. Towards a post-modern reconstruction of ethics. *University of Pennsylvania Law Review*, 133: 291–380.

Cornish W., 1999. *Intellectual property: patents, copyright, trademarks and allied rights*. Sweet and Maxwell, London.

Correa C., 1995. Sovereign and property rights over plant genetic resources. *Agriculture and Human Values*, 12/4: 58–79.

Cosgrove D., 1990. Environmental thought and action: pre-modern and post-modern. *Transactions of the Institute of British Geographers*, 15/3: 344–58.

Cotterill F., 1995. Systematics, biological knowledge and environmental conservation. *Biodiversity and Conservation*, 4: 183–205.

Cox S., 1985. No tragedy of the commons. *Environmental Ethics*, 7: 49–61.

(CPGR) Commission on Plant Genetic Resources, 1987a. *Legal status of base and active collections of PGR*. Item 5, Provisional agenda, 2nd session (16–20 March). CPGR/87/5: 4–9. Food and Agriculture Organization, Rome.

(CPGR) Commission on Plant Genetic Resources, 1987b. *Study on legal arrangements with a view to the possible establishment of an international network of base collections in genebanks under the auspices or jurisdiction of FAO*. Item 6, Provisional agenda, 2nd session (16–20 March). CPGR/87/6. FAO, Rome.

(CPGR) Commission on Plant Genetic Resources, 1993. *Progress report on the global system for the conservation and utilisation of plant genetic resources*. 5th session of the CPGR (19–23) April. CPGR/93/5. FAO, Rome.

(CPGR) Commission on Plant Genetic Resources, 1994a. *The international network of* ex situ *germplasm collections: progress report*. 9th session of Working Group of the CPGR (11–12 May). CPGR/94/WG9/6. FAO, Rome.

(CPGR) Commission on Plant Genetic Resources, 1994b. *Survey of existing data on* ex situ *collections of PGR for food and agriculture*. 1st extraordinary session of the CPGR (7–11 November). CPGR–Ex1/94/5/Annex. FAO, Rome.

(CPGR) Commission on Plant Genetic Resources, 1994c. *Revision of the International Undertaking. Stage 1: integration of the annexes and harmonization with the Convention on Biological Diversity.* 1st extraordinary session of the CPGR (7–11 November). CPGR–Ex1/94/4/Alt. FAO, Rome.

Cracraft J., 1995. The urgency of building global capacity for biodiversity science. *Biodiversity and Conservation*, 4: 463–75.

Crang M. and N. Thrift (eds), 2000. *Thinking space*. Routledge, London.

Crawford T., 1993. An interview with Bruno Latour. *Configurations*, 1/2: 247–268.

Crocodile Specialist Group, 1997a. *The story of the crocodile specialist group.* http://www.flmnh.ufl.edu/natsci/herpetology/crocs/crocsb.htm

Crocodile Specialist Group, 1997b. *Crocodile action plan 08.* http://www.flmnh.ufl.edu/natsci/herpetology/crocs/crocsb.htm

Cronon W., 1983. *Changes in the land: Indians, colonists and the ecology of New England.* Hill and Wang, New York.

Cronon W., 1995. The trouble with wilderness, or getting back to the wrong nature. In *Uncommon ground*: 69–90. W.W. Norton & Co., New York.

Cronon W. (ed.), 1995. *Uncommon ground. Toward reinventing nature.* W.W. Norton & Co., New York.

Crucible Group (The), 1994. *People, plants and patents.* International Development Research Center, Ottawa, Canada.

Cunningham H., 1998. Colonial encounters in post-colonial contexts: patenting indigenous DNA and the Human Genome project. *Critique of Anthropology*, 18: 205–33.

Curtin D., 1991. Towards an ecological ethic of care. *Hypatia*, 6/1: 60–74.

Cussins C., 1997. Elephants, biodiversity and complexity: Amboseli National Park, Kenya. Unpublished manuscript presented at Actor Network Theory conference, University of Lancaster, September.

Davies G., 1999. Exploiting the archive: and the animals came in two by two, 16mm, CD-ROM and BetaSp. *Area*, 31/1: 49–58.

Delaney D., 2001. Making nature/marking humans: law as a site of (cultural) production. *Annals of the Association of American Geographers*, 91/3: 487–503.

Delannay X. *et al.* (32 co-authors), 1991. Yield evaluation of a glyphosate resistant soybean line after treatment with glyphosate. *Crop Science*, 35: 1461–7.

Deleuze G., 1990 (1969 in French). *The logic of sense* (trans. M. Lester and C. Stivale). Athlone Press, London.

Deleuze G., 1993. Rhizome versus trees. In Boundas C. (ed.), *The Deleuze reader*, Columbia University Press, New York.

Deleuze G., 1994 (1968 in French). *Difference and repetition* (trans. P. Patton). Athlone Press, London.

Deleuze G., 1995 (1990 in French). *Negotiations* (trans. M. Joughin). Columbia University Press, New York.

Deleuze G. and F. Guattari, 1988 (1980 in French). *A thousand plateaus. Capitalism and schizophrenia* (trans. B. Massumi). Athlone Press, London.

Demeritt D., 1998. Science, social constructivism and nature. In Castree N. and B. Willems-Braun (eds), *Remaking reality. Nature at the millennium*: 173–93. Routledge, London.

der Derian J. and M. Shapiro (eds), 1989. *International/intertextual relations. Postmodern readings of world politics.* Lexington Books, Lexington, MA.

Derrida J., 1991. 'Eating well', or the calculation of the subject. An interview with Jacques Derrida. In Cadava E., P. Connor, and J.-L. Nancy (eds), *Who comes after the subject*: 96–119. Routledge, London.

Descola P. and G. Palsson (eds), 1996. *Nature and society: anthropological perspectives*. Routledge, London.

Dickens P., 1996. *Reconstructing nature*. Routledge, London.

Diprose R., 1994. *The bodies of women. Ethics, embodiment and sexual difference*. Routledge, London.

Dobson A. and D. Lucardie (eds), 1995. *The politics of nature*. Routledge, London.

Dodson M., 1994. Towards the exercise of indigenous rights: policy, power and self-determination. *Race and Class*, 35/4: 65–76.

Dolins F. (ed.), 1999. *Attitudes to animals. Views on animal welfare*. Cambridge University Press, Cambridge.

Donkin R., 1989. *The Muscovy duck*. A.A. Balkema, Rotterdam.

Donovan J., 1993. Animal rights and feminist theory. In Gaard G. (ed.), *Ecofeminism. Women, animals, nature*: 167–94. Temple University Press, Philadelphia, PA.

Douglas M., 1966. *Purity and danger*. Routledge & Kegan Paul, London.

Doyle R., 1997. *On beyond living. Rhetorical transformations of the life sciences*. Stanford University Press, Stanford, CA.

Drahos P., 1996. *A philosophy of intellectual property*. Dartmouth Press, Aldershot.

Drayton R., 2000. *Nature's government. Science, imperial Britain, and the 'improvement' of the world*. Yale University Press, New Haven, CT.

Driver F., 1985. Power, space and the body: a critical assessment of Foucault's *Discipline and punish*. *Society and Space*, 10: 23–40.

Driver F., 1992. Geography's empire: histories of geographical knowledge. *Society and Space*: 10: 23–40.

Dryzek J., 1990. Green reason: communicative ethics for the biosphere. *Environmental Ethics*, 12: 195–210.

Durant J., 1998. Written evidence to the House of Lords Select Committee on European Communities. *Second report on the regulation of genetic modification in agriculture*. http://www.publications.parliament.uk/pa/ld199899/ldselect/ldeucom/11/8121501.htm

(EAC) Environmental Audit Committee, 1999. *Genetically Modified Organisms and the environment: co-ordination of Government policy*. Vol. 1: Report and proceedings, May 1999; Vol. 2: Evidence, May 1999. House of Commons, HC 384 – I and II.

Earnhardt J., S. Thompson and K. Willis, 1995. ISIS database: an evaluation of records essential for captive management. *Zoo Biology*, 14: 493–508.

Earthwatch, 1997. *Expeditions Brochure*. Earthwatch, Boston, MA.

Earthwatch Institute, 1998. http://www.earthwatch.org

Ebert T., 1991. The (body) politics of feminist theory. *Phoebe*, 3/2: 56–65.

Edgeworth B., 1994. Tenure, allodialism and indigenous rights at common law: English, United States and Australian land law compared after *Mabo v. Queensland*. *The Anglo-American Law Review*, 23 (Oct/Dec): 397–434.

Elam M., 1999. Living dangerously with Bruno Latour in a hybrid world. *Theory, Culture and Society*, 16/4: 1–24.

Elder G., J. Wolch and J. Emel, 1998. Race, place and the bounds of humanity. *Society and Animals*, 6/2: 183–202.

Eldridge N., 1995. *Reinventing Darwin*. John Wiley, New York.

(ENDS) Environmental Data Services, 1999. Sainsburys and M&S in 'GM-free' retailer consortium. *ENDS report*, 290: 33.

English Nature, 1998. Written evidence to the House of Lords Select Committee on European Communities. *Second report on the regulation of genetic modification in agriculture.* www.publications.parliament.uk/pa/ld199899/ldselect/ldeucom/11/11we13

Environmental History, 1997. Commentaries on Cronon's essay 'The trouble with wilderness', and author's response. *Environmental History*, 1/1: 29–55.

Epstein R., 1979. Possession as the root of title. *Georgia Law Review*, 13: 1197–243.

Escobar A., 1995. *Encountering development. The making and unmaking of the Third World.* Princeton University Press, Princeton, NJ.

(ESRC) Economic and Social Research Council, 1999. *The politics of GM food. Risk science and public trust.* ESRC Global Environmental Change Programme Special Briefing no. 5.

European Community, 1998. Council regulation 1139/98 concerning the compulsory indication on the labelling of certain foodstuffs produced from genetically modified organisms. *Official Journal of the European Communities*, L159, 3 June: 4.

Evans, L. 1998.*Feeding the ten billion: plants and population growth*. Cambridge University Press, Cambridge.

Fallon S. and M. Enig, 2000. Tragedy and hype: the third Soy symposium. www.nexusmagazine.com

(FAO) Food and Agriculture Organization, 1979. *Verbatim proceedings of the 8th meeting of Commission II.* 20th session of FAO conference (10–28 November). C/79/II/PV/8: 177). FAO, Rome.

(FAO) Food and Agriculture Organization, 1983a. *Report of 22nd session of FAO Conference* (5–23 November). C/83/rep. Annex to resolution 8/83. FAO, Rome.

(FAO) Food and Agriculture Organization, 1983b. *Verbatim proceedings of the 15th meeting of Commission II.* 22nd session of FAO conference (5–23 November). C/83/II/PV/15. FAO, Rome.

(FAO) Food and Agriculture Organization, 1983c. *Proposal for the establishment of an international genebank and preparation of a draft international treaty for PGR.* 7th session of Committee on Agriculture (21–30 March). CAOG/83/10. FAO, Rome.

(FAO) Food and Agriculture Organization, 1985a. *Verbatim proceedings of the 11th meeting of Commission II.* 23rd session of FAO conference (9–28 November). C/85/II/PV/11. FAO, Rome.

(FAO) Food and Agriculture Organization, 1985b. *Verbatim proceedings of the 12th meeting of Commission II.* 23rd session of FAO conference (9–28 November). C/85/II/PV/11. FAO, Rome.

(FAO) Food and Agriculture Organization, 1988. *Report of the 88th session of the FAO Council* (4–7 November). CL/88/rep/1. FAO, Rome.

(FAO) Food and Agriculture Organization, 1989a. *Report of the 25th session of the FAO Conference* (11–30 November). C/89/rep. FAO, Rome.

(FAO) Food and Agriculture Organization, 1989b. *Verbatim proceedings of the 8th meeting of Commission I.* 25th session of FAO Conference (11–29 November). C/89/I/PV/8. FAO, Rome.

(FAO) Food and Agriculture Organization, 1991a. *Report of the 26th session of FAO conference* (9–27 November). C/91/rep. FAO, Rome.

(FAO) Food and Agriculture Organization, 1991b. *Verbatim proceedings of the 8th meeting of Commission I.* 26th session of FAO Conference (9–27 November). C/91/I/PV/8. FAO, Rome.

(FAO) Food and Agriculture Organization, 1993. *Harvesting nature's diversity.* Information Division, FAO, Rome.

FAO/WHO (World Health Organization), 1996. *Joint expert consultation on Biotechnology and food safety.* FAO, Rome.

Ferguson M., 1992. Mary Wollstonecraft and the problematic of slavery. *Feminist Review*, 82: 102–24.

Ferry L., 1992. *Le nouvel ordre écologique.* Grasset, Paris.

Fischler F., 1997. Genetically modified or not? The answer is the consumers. www.europa.int/en/comm/dg06/com/htmfiles/welcome.htm

Fiddes N., 1991. *Meat. A natural symbol.* Routledge, London.

Fine B., M. Heasman and J. Wright, 1996. *Consumption in the age of affluence: the world of food.* Routledge, London.

Finkielkraut A., 2001 (1986 in French) *In the name of humanity.* Pimlico, London.

Fitzpatrick P., 1992. *The mythology of modern law.* Routledge, London.

Fitzsimmons M., 1989. The matter of nature. *Antipode*, 21/2: 106–20.

Fitzsimmons M. and D. Goodman, 1998. Incorporating nature: environmental narratives and the reproduction of food. In Castree N. and B. Willems-Braun (eds), *Remaking reality. Nature at the millennium*: 194–220. Routledge, London.

Flitner M., 1998. Biodiversity: local commons or global commodities. In Goldman M. (ed.), *Privatizing nature*: 144–66. Pluto Press, London.

Food and Drink Federation, 1998. *Supplementary memorandum submitted in evidence to the House of Lords Select Committee on European Communities Second report on Genetic Modification.* http://www.publications.parliament.uk/pa/ld199899/ldselect/ldeucom/11/11we57.htm

Ford B., 2000. *The future of food.* Thames and Hudson, London.

Foreman D., 1981. Earth First! *The Progressive*, 45: 40–42. Reprinted in List P. (ed.), 1993. *Radical environmentalism*: 187–92. Wadsworth, Belmont, CA.

Foucault M., 1973. *The order of things.* Vintage, New York.

Foucault M., 1986. Of other spaces. *Diacritics*, 16/1: 22–7.

Fowler C. and P. Mooney, 1990. *Shattering. Food politics, and the loss of genetic diversity.* The University of Arizona Press, Tucson, AZ.

Fox W., 1990. *Towards a transpersonal ecology.* Shambhala Publications, London.

Frankel O. and E. Bennett (eds), 1970. *Genetic resources in plants.* International Biological Programme, London.

Freeman R., 1972. *Classification of the animal kingdom: an illustrated introduction.* Hodder and Stoughton, Sevenoaks, Kent.

Friedland W., A. Barton and R. Thomas, 1981. *Manufacturing green gold: capital, labour and technology in the lettuce industry.* Cambridge University Press, Cambridge.

Friedman M., 1989. Feminism and modern friendship: dislocating the community. *Ethics*, 99: 275–90.

Friis-Hansen E., 1994. Conceptualising *in situ* conservation of landraces. In Krattiger A (ed.), *Widening perspectives on biodiversity*: 263–276. IUCN, Switzerland.

Frow J., 2001. The politics of stolen time. In May J. and N. Thrift (eds), *Timespace. Geographies of Temporality*: 38–56. Routledge, London.

Fuss D., 1989. *Essentially Speaking: Feminism, nature and difference*. Routledge, London.

Futrell A., 1997. *Blood in the arena: the spectacle of Roman power*. University of Texas Press, Austin, TX.

Game A. and A. Metcalfe, 1996. *Passionate sociology*. Sage, London.

Gannon P., T. Guthrie and G. Laurie, 1995. Patents, morality and DNA: should there be intellectual property protection of the human genome project? *Medical Law International*, 1: 321–45.

Gelder K. and J. Jacobs, 1995. Uncanny Australia. *Ecumene*, 2: 171–83.

Gelder K. and J. Jacobs, 1998. *Uncanny Australia. Sacredness and identity in a postcolonial nation*. University of Melbourne Press, Melbourne.

George W., 1962. *Animal Geography*. Heinemann, London.

Gibson R., 1992. *South of the west. Post colonialism and the narrative construction of Australia*. Indiana University Press, Bloomington, IN.

Gibson-Graham J.K., 1996. *The end of capitalism (as we knew it)*. Basil Blackwell, Oxford.

Giddens A., 1991. *Modernity and self-identity*. Polity Press, Oxford.

Gilligan C., 1982. *In a different voice*. Harvard University Press, Cambridge, MA.

Glacken C., 1967. *Traces on the Rhodian Shore. Nature and culture in western thought from ancient times to the end of the eighteenth century*. University of California Press, Berkeley, CA.

Glendinning S., 1998. *On being with others*. Routledge, London.

Gobetti D., 1992. *Private and public. Individuals, households and body politic in Locke and Hutcheson*. Routledge, London.

Goldman M. (ed.), 1998. *Privatizing nature: political struggles for the global commons*. Pluto Press, London.

Goodman D., 1999. Agro-food studies in the 'age of ecology': nature, corporeality, bio-politics. *Sociologia Ruralis*, 39/1: 17–38.

Goodman D., A. Sorj and J. Wilkinson, 1987. *From farming to biotechnology: a theory of agro-industrial development*. Basil Blackwell, Oxford.

Goodrich P., 1991. Eating law: commons, common land, common law. *Journal of Legal History*, 12: 246–63.

Goodrich P., C. Douzinas and Y. Hachamovitch, 1994. Politics, ethics and the legality of the contingent. In Douzinas C., P. Goodrich and Y. Hachamovitch (eds), *Politics, postmodernism and critical legal studies*: 1–34. Routledge, London.

Goodwin B., 1994. *How the leopard changed its spots*. Phoenix, London.

Gore A., 1992. *Earth in balance: ecology and the human spirit*. Houghton Miffin, Boston, MA.

Gottlieb R. (ed.), 1997. *The ecological community*. Routledge, London.

Gottwiess H., 1998. *Governing molecules. The discursive politics of genetic engineering in Europe and the United States*. The MIT Press, Cambridge, MA.

Graves-Brown P. (ed.), 2000. *Matter, materiality and modern culture*. Routledge, London.

Greenblatt S., 1991. *Marvellous possessions. The wonder of the new world*. University of Chicago Press, Chicago, IL.

Greene K., 1986. *The archaeology of the Roman economy*. Batsford, London.

Gregory D., 1994. *Geographical imaginations*. Basil Blackwell, Oxford.

Griffin D., 1992. *Animal minds*. Chicago University Press, Chicago, IL.

Griffiths S. and J. Wallace (eds), 1998. *Consuming passions: food in the age of anxiety*. Mandolin, Manchester.

Gross E., 1986. Philosophy, subjectivity and the body. Kristeva and Irigaray. In Pateman C. and E. Gross (eds), *Feminist Challenges*: 125–43. Allen and Unwin, Sydney.

Gross P. and N. Levitt, 1994. *Higher superstition: the academic left and its quarrels with science*. Johns Hopkins University Press, Baltimore, MD.

Grosz E., 1989. *Sexual subversions*. Allen and Unwin, Sydney.

Grosz E., 1994. *Volatile bodies: toward a corporeal feminism*. Indiana University Press, Bloomington, IN.

Grove R., 1995. *Green imperialism. Colonial expansion, tropical island edens and the origins of environmentalism 1600–1860*. Cambridge University Press, Cambridge.

Grove-White R., P. Macnaghten, S. Mayer and B. Wynne, 1997. *Uncertain world: Genetically Modified organisms, food and public attitudes in Britain*. Centre for the Study of Environmental Change, Lancaster University, Lancaster.

Grusin R., 1998. Reproducing Yosemite: Olmsted, environmentalism, and the nature of aesthetic agency. *Cultural Studies*, 12/3: 332–59.

Haila Y., 1997. 'Wilderness' and the multiple layers of environmental thought. *Environment and History*, 3/2: 129–48.

Halberstam J. and I. Livingston (eds), 1995. *Posthuman bodies*. Indiana University Press, Bloomington, IN.

Ham J., 1997. Taming the beast: animality in Wedekind and Nietzsche. In Ham J. and M. Senior (eds), *Animal acts*: 145–64. Routledge, London.

Ham J. and M. Senior (eds), 1997. *Animal acts. Configuring the human in western history*. Routledge, London.

Hamilton N., 1993. Who owns dinner? Evolving legal mechanisms for ownership of plant genetic resources. *Tulsa Law Journal*, 28: 587–657.

Hampson F. and J. Reppy (eds), 1996. *Earthly goods. Environmental change and social justice*. Cornell University Press, Ithaca, NY.

Hancock J., 1992. *Plant evolution and the origin of crop species*. Prentice-Hall, Englewood Cliffs, NJ.

Hann C. (ed.), 1998. *Property relations. Renewing the anthropological tradition*. Cambridge University Press, Cambridge.

Hannigan J., 1995. *Environmental sociology*. Routledge, London.

Hansard, 1993a. Native Title Bill. Commonwealth of Australia Parliamentary debates. *Weekly Hansard, House of Representatives*. (16–18 & 22–25 November). Commonwealth Government Printer, Canberra, Australia.

Hansard, 1993b. Native Title Bill. Commonwealth of Australia Parliamentary debates. *Weekly Hansard, Senate*. (6–9 November). Commonwealth Government Printer, Canberra, Australia.

Hansard, 1993c. Native Title Bill. Commonwealth of Australia Parliamentary debates. *Daily Hansards, Senate.* (14–17 & 20 December). Commonwealth Government Printer, Canberra, Australia.

Hansen M., 2000a. *Embodying technesis. Technology beyond writing.* University of Michigan Press, Michigan, MI.

Hansen M., 2000b. Becoming as creative involution? Contextualizing Deleuze and Guattari's biophilosophy. *Postmodern Culture,* 11/1: 1–42.

Hapgood J., 1987. Soybean. *National Geographic,* 172: 67–91.

Haraway D., 1985. Manifesto for cyborgs: science, technology and socialist feminism in the 1980s. *Socialist review,* 80: 65–108.

Haraway D., 1989. *Primate visions. Gender, race and nature in the world of modern science.* Routledge Chapman and Hall, London.

Haraway D., 1991a. Situated knowledges: the science question in feminism and the privilege of partial perspective. In *Simians, cyborgs and women. The reinvention of Nature:* 183–202. Free Association Books, San Francisco, CA.

Haraway D., 1991b. *Simians, cyborgs, and women. The reinvention of nature.* Free Association Books, London.

Haraway D., 1992. Otherworldly conversations; terran topics; local terms. *Science as Culture,* 3/1: 64–98.

Haraway D., 1993. The promises of Monsters: a regenerative politics for Inappropriate/d Others. In Grossberg L., C. Nelson and P. Treichler (eds), *Cultural Studies:* 295–337. Routledge, London.

Haraway D., 1995. Nature, politics and possibilities: a debate and discussion with David Harvey and Donna Haraway. *Society and Space,* 13: 507–27.

Haraway D., 1997. *Modest witness@ second millennium. FemaleMan meets Onco-Mouse.* Routledge, London.

Haraway D., 2000. *How like a leaf. An interview with T. Goodeve.* Routledge, London.

Hardin G., 1968. The tragedy of the commons. *Science,* 162: 1243–8.

Hargreaves I., 2000. *Who's misunderstanding whom: science, society and the media?* ESRC, Swindon. http://www.esrc.ac.uk

Hardon J., B. Vosman and Th. Van Hintum, 1994. *Identifying genetic resources and their origin: the capabilities of modern biochemical and legal systems.* CPGR, Background study paper no. 4. FAO, Rome.

Hardt M., 1993. *An apprenticeship in philosophy. Gilles Deleuze.* University College London Press, London.

Harlan J., 1971. Agricultural origins: centers and noncenters. *Science,* 174: 468–74.

Harlan J., 1975. Our vanishing genetic resources. *Science,* 188: 618–21.

Harvey D., 1996. *Justice, nature and the geography of difference.* Basil Blackwell, Oxford.

Hayles K.N., 1995. Searching for common ground. In Soulé M. and G. Lease (eds), *Reinventing nature?:* 47–64. Island Press, Washington, DC.

Hayles K.N., 1999. *How we became post-human.* University of Chicago Press, Chicago, IL.

Helgason A. and G. Pálsson, 1997. Contested commodities: the moral landscape of modernist regimes. *Journal of the Royal Anthropological Institute,* 3: 451–71.

Hennessy R., 1993. *Materialist feminism and the politics of discourse.* Routledge, London.

Hetherington K., 1997a. *The badlands of modernity: heterotopias and social ordering*. Routledge, London.

Hetherington K., 1997b. Museum topology, the will to connect. *Journal of Material Culture*, 2/2: 199–218.

Hetherington K., 1997c. In place of geometry: the materiality of place. In Hetherington K. and R. Munro (eds), *Ideas of difference: social spaces and the labour of division*: 183–99. Basil Blackwell, Oxford.

Hetherington K. and N. Lee, 2000. Social order and the blank figure. *Society and Space*, 18: 169–84.

Hinchee M., D. Connor-Ward, C. Newell, R. McDonnell, S. Sato, C. Gasser, D. Fischoff, D. Re, R. Fraley and B. Horsch, 1988. Production of Transgenic soybean plants using *Agrobacterium*-mediated DNA transfer. *Bio/Technology*, 6: 915–22.

Hinchliffe S., 1999. Entangled humans. In Philo C., P. Routledge and J. Sharp (eds), *Entanglements of power*: 219–7. Routledge, London.

Hinchliffe S., 2001. Indeterminacy indecisions – science, policy and politics in the BSE crisis. *Transactions of the Institute of British Geographers*, 26/2: 182–204.

Hirsch A. and T. LaRue, 1997. Is the legume nodule a modified root or stem or an organ *sui generis*? *Critical Reviews in Plant Sciences*, 16/4: 361–92.

Ho M.-W., 1999 (2nd edition). *Genetic engineering. Dream or nightmare?* Gateway, Dublin.

Ho M.-W. and R. Steinbrecher, 1998. Fatal flaws in food safety assessment. Critique of the Joint FAO–WHO Biotechnology and Food Safety Report. *Environmental and Nutritional Interactions*, 2: 51–84.

Hobbes T., 1985 (1st published 1651). *Leviathan*. Penguin, Harmondsworth.

Hobhouse H., 1999 (2nd edition). *Seeds of change. Six seeds that transformed mankind*. Macmillan, London.

Hoeg P., 1996. *The woman and the ape*. Picador, London.

Honoré A.M., 1961. Ownership. In Guest A.G. (ed.), *Oxford essays in jurisprudence: first series*. Oxford University Press, Oxford.

hooks b., 1990. Third world diva girls. In *Yearning: race, gender and cultural politics*: 89–102. Turnaround, London.

House of Lords Select Committee on European Communities, 1999. *Second report on the regulation of genetic modification in agriculture*. www.publications. parliament.uk/pa/ld199899/ldselect/ldeucom

Hughes J., 1994. *Pan's travail. Environmental problems of the ancient Greeks and Romans*. Johns Hopkins University Press, Baltimore, MD.

Hulme P., 1990. The spontaneous hand of nature. In Hulme P. and L. Jordanova (eds), *The enlightenment and its shadows*: 18–32. Routledge, London.

Hymowitz T. and J. Harlan, 1983. Introduction of soybean to North America by Samuel Bowen in 1765. *Economic Botany*, 37/4: 373.

Ignatieff M., 1984. *The needs of strangers*. Chatto and Windus, London.

ILDIS, 2000. *International Legume Database and Information Service website* http://www/ildis.org/

Ingold T., 1986. *The appropriation of nature. Essays on human ecology and social relations*. Manchester University Press, Manchester.

Ingold T. (ed.), 1988a. Introduction to *What is an animal?* Routledge, London.

Ingold T., 1988b. The animal in the study of humanity. In T. Ingold (ed.), *What is an animal?*: 84–99. Routledge, London.

Ingold T., 1989. An anthropologist looks at biology. *Man* (N.S.), 25: 208–29.

Ingold T., 1993. Globes and spheres: the topology of environmentalism. In Milton K. (ed.), *Environmentalism. The view from anthropology*: 31–42. Routledge, London.

Ingold T., 1995a. Building, dwelling, living: how animals and people make themselves at home in the world. In Strathern M. (ed.), *Shifting contexts. Transformations in anthropological knowledge*: 57–80. Routledge, London.

Ingold T., 1995b. 'People like us': the concept of the anatomically modern human. *Cultural Dynamics*, 7/2: 187–214.

Ingold T., 2000a. *The perception of the environment. Essays in livelihood, dwelling and skill*. Routledge, London.

Ingold T., 2000b. Making culture and weaving the world. In Graves-Brown P. (ed.), *Matter, materiality and modern culture*: 50–71. Routledge, London.

Institute of Food Research, 2000. *Information sheet on soya*. www.ifrn.bbsrc.ac.uk/public.FoodInfoSheets/soya.html

Irigaray L., 1999. *The forgetting of air (in Martin Heidegger)* (trans. M. Mader). Athlone Press, London.

ISAAA, 1998. Global review of commercialised transgenic crops. ISAAA Briefing no. 8 (author C. James). Ithaca, NY.

Jackson P., 1999. Commodity cultures: the traffic in things. *Transactions of the Institute of British Geographers*, 24/1: 95–108.

Jacobs J., 1995. *Edge of empire*. Routledge, London.

Jardine N., J. Secord and E. Spary (eds), 1996. *Cultures of natural history*. Cambridge University Press, Cambridge.

Jeffries M., 1997. *Biodiversity and conservation*. Routledge, London.

Jenkins M. and S. Broad, 1994. *International trade in reptile skins: a review and analysis of the main consumer markets, 1983–91*. Traffic International, Cambridge.

Jennison G., 1937. *Animals for show and pleasure in ancient Rome*. Manchester University Press, Manchester.

Johnson G. and M. Smith (eds), 1990. *Ontology and alterity in Merleau-Ponty*. Northwestern University Press, Evanston, IL.

Jones G. (ed.), 1972. Introduction to *The sovereignity of the law. Selections from Blackstone's commentaries on the laws of England*. Macmillan, London.

Juma C., 1989. *The gene hunters. Biotechnology and the scramble for seeds*. Zed Books, London.

Kafka F., 1988 (1905). *Metamorphosis*. Penguin, Harmondsworth.

Kaufman L. and K. Mallory (eds), 1993 (2nd edition). *The last extinction*. The MIT Press, Cambridge, MA.

Keller E. Fox, 1983. *A feeling for the organism: the life and work of Barbara McClintock*. W.H. Freeman and Co., San Francisco, CA.

Kelley D., 1990. *The human measure*. Harvard University Press, Cambridge, MA.

Kendall G. and G. Wickham, 1999. *Using Foucault's methods*. Sage, London.

Kenney M., 1986. *Biotechnology: the university–industrial complex*. Yale University Press, New Haven, CT.

Keon-Cohen B., 1993. Some problems of proof: the admissability of traditional evidence. In Stephenson M. and S. Ratnapala (eds), *Mabo: a judicial revolution*: 185–205. University of Queensland Press, St Lucia, Queensland.

Kerridge R. and N. Samuells (eds), 1998. *Writing the environment*. Zed Books, London.

King R., 1991. Caring about nature: feminist ethics and the environment. *Hypatia*, 6/1: 75–89.

Kingdon J., 1997. *The Kingdon fieldguide to African Mammals*: 504–8. Academic Press, London.

Kiple K. and K. Ornelas (eds), 2000. *The Cambridge world history of food*. Cambridge University Press, Cambridge.

Kirkby V., 1997. *Telling flesh. The substance of the corporeal*. Routledge, London.

Kloppenburg J., 1988. *First the seed. The political economy of plant biotechnology 1492-2000*. Cambridge University Press, Cambridge.

Kloppenburg J., 1995. *Wither farmers' rights?* A consultancy report for the Commission for Plant Genetic Resources. Mimeo.

Kloppenburg J. and D. Kleinman, 1987a. Seed wars: common heritage, private property and political strategy. *Socialist Review*, 95: 7–41.

Kloppenburg J. and D. Kleinman, 1987b. The plant germplasm controversy: analysing empirically the distribution of the world's plant genetic resources. *BioScience*, 37: 190–8.

Koerner L., 1996. Carl Linnaeus in his time and place. In Jardine N., J. Secord and E. Spary (eds), *Cultures of natural history*: 145–62. Cambridge University Press, Cambridge.

Kotz S., 1976. The common heritage of mankind. Resource management of the international seabed. *Ecology Law Review*, 6: 65–108.

Krebs J. and A. Kacelnick, 1997. Risk: a scientific view. In *Science, risk and precaution*: 31–44. Royal Society, London.

Kruks S., 1995. Identity politics and dialectical reason: beyond an epistemology of provenance. *Hypatia*, 10/2: 1–22.

Kuper A., 1988. *The invention of primitive society. Transformations of an illusion*. Routledge, London.

Kymlicka W., 1991. The social contract tradition. In Singer P. (ed.), *A companion to ethics*: 186–96. Basil Blackwell, Oxford.

Laclau E. and C. Mouffe, 1985. *Hegemony and socialist strategy. Towards a radical democratic politics*. Verso, London.

de Landa M., 1997. *A thousand years of nonlinear history*. Zone Books, New York.

de Landa M., 1999. Deleuze, diagrams and open-ended becoming. In Grosz E. (ed.), *Becomings*: 29–41. Routledge, London.

Langer M., 1990. Merleau-Ponty and deep ecology. In Johnson G. and M. Smith (eds), *Ontology and alterity in Merleau-Ponty*: 115–29. Northwestern University Press, Evanston, IL.

Lappé M. and B. Bailey, 1999. *Against the grain. The genetic transformation of global agriculture*. Earthscan, London.

Larschan B. and B. Brennan, 1983. The common heritage of mankind principle in international law. *Columbia Journal of Transnational Law*, 21: 305–37.

Latour B., 1988. *The pasteurization of France*. Harvard University Press, Cambridge, MA.

Latour B., 1993. *We have never been modern* (trans. C. Porter). Harvester Wheatsheaf, Hemel Hempstead.

Latour B., 1994a. Pragmatologies. *American Behavioural Scientist*, 37/6: 791–808.

Latour B., 1994b. On technical mediation – philosophy, sociology, genealogy. *Common Knowledge*, 3/2: 29–64.

Latour B., 1996. *Aramis or the love of technology* (trans. C. Porter). Harvard University Press, Cambridge, MA.

Latour B., 1997a. Trains of thought: Piaget, formalism and the fifth dimension. *Common Knowledge*, 6/3: 170–91.

Latour B., 1997b. Foreword to *Power and Invention* (I. Stengers): vii–xx. University of Minnesota Press, Minneapolis, MN.

Latour B., 1999a. *Pandora's hope. Essays on the reality of science studies.* Harvard University Press, Cambridge, MA.

Latour B., 1999b. On recalling ANT. In Law. J. and J. Hassard (eds), *Actor network theory and after*: 15–25. Basil Blackwell, Oxford.

Latour B., 1999c. *Politiques de la nature. Comment faire entrer les sciences en démocratie.* Armillaire, Paris.

Latour B., 2000. When things strike back. *British Journal of Sociology*, 51/1: 107–23.

Law J., 1986. On methods of long-distance control: vessels, navigation and the Portuguese route to India. *Sociological Review Monograph* 32: 234–63.

Law J. (ed.), 1991. *A sociology of monsters: essays on power, technology and domination.* Routledge, London.

Law J., 1994. *Organizing modernity.* Basil Blackwell, Oxford.

Law J. and J. Hassard (eds), 1999. *Actor network theory and after.* Basil Blackwell, Oxford.

Law J. and A. Mol, 1995. Notes on materiality and sociality. *Sociological Review*, 42/3: 274–94.

Lawrence G., 1990. Agricultural restructuring and rural change in Australia. In Marsden T., P. Lowe and S. Whatmore (eds), *Restructuring rurality*: 101–29. John Wiley, Chichester.

Lear L., 1997. *Rachel Carson. The life of the author of Silent Spring.* Allen Lane, The Penguin Press, London.

Leder D., 1990a. Flesh and blood: a proposed supplement to Merleau-Ponty. *Human Studies*, 13: 209–19.

Leder D., 1990b. *The absent body.* University of Chicago Press, Chicago, IL.

Leighly J. (ed.), 1963. *Land and life: selections from the writings of Carl Ortwin Sauer.* University of California Press, Berkeley, CA.

Levidow L., 1999. Britain's biotechology controversy: elusive science, contested expertise. *New Genetics and Society*, 18/1: 47–64.

Levidow L. and S. Carr, 1996. UK biotechnology regulation: disputing regulatory boundaries. *Science and Public Policy*, 23/3: 164–70.

Levin D., 1990. Justice in the flesh. In Johnson G. and M. Smith (eds), *Ontology and alterity in Merleau-Ponty*: 35–44. Northwestern University Press, Evanston, IL.

Lewontin R., 1998. *The triple helix: gene, organism and environment.* Harvard University Press, Cambridge, MA.

Lines W., 1991. *Taming the great south land. A history of the conquest of nature in Australia.* University of California Press, Berkeley CA.

Lipietz A., 1995. Enclosing the global commons: global environmental negotiations in a North-South conflictual approach. In Bhaskar R. and A. Glynn (eds) *The North, The South and the Environment.* Earthscan, London.

Lipschutz R., 1998. The nature of sovereignty and the sovereignty of nature: problematizing the boundaries of self, society, state and system. In Litfin K. (ed.), *The greening of sovereignty in world politics*: 109–40. The MIT Press, Cambridge, MA.

Litfin K. (ed.), 1998. *The greening of sovereignty in world politics*. The MIT Press, Cambridge, MA.

Livingstone D., 1992. *The geographical tradition*. Basil Blackwell, Oxford.

Livingstone D., 1998. Reproduction, representation and authenticity: a re-reading. *Transactions of the Institute of British Geographers*, 23: 13–19.

Locke J., 1988 (1st published 1690). *Two treatises on government*. Cambridge University Press (student edition), Cambridge.

Lorraine D., 1999. *Irigaray and Deleuze: experiments in visceral philosophy*. Cornell University Press, Ithaca, NY.

Lovibond S., 1994. Maternalist ethics: a feminist reassessment. *The South Atlantic Quarterly*, 93/4: 779–802.

Luke B., 1997. Solidarity across diversity. A pluralistic rapprochement of environmentalism and animal liberation. In Gottlieb R. (ed.), *The ecological community*: 33–358. Routledge, London.

Luke T., 1996. Liberal society and cyborg subjectivity: the politics of environments, bodies, and nature. *Alternatives*, 21: 1–30.

Luke T., 1997. *Eco-critique. Contesting the politics of nature, economy, and culture*. University of Minnesota Press, Minneapolis, MN.

Lupton D., 1996. *Food, the body and the self*. Sage, London.

Lyster S., 1985. *International wildlife law: an analysis of international treaties concerned with the conservation of wildlife*. Grotius Publications, Llandysul, Wales.

Macauley D. (ed.), 1996. *Minding nature: the philosophers of ecology*. Guilford Press, New York.

Macauley D., 1997. Be-wildering order: on finding a home for domestication and the domesticated other. In Gottlieb R. (ed.), *The ecological community*: 105–31. Routledge, London.

Macfarlane A., 1998. The mystery of property: inheritance and industrialization in England and Japan. In Hann (ed.), *Property relations*: 104–23. Cambridge University Press, Cambridge.

Macgregor R., 1993. A doomed race: a scientific axiom of the late nineteenth century. *Australian Journal of Politics and History*, 39/1: 14–22.

Macnaghten P. and J. Urry, 1999. *Contested natures*. Sage, London.

Macnaghten P. and J. Urry (eds), 2000. Bodies of nature. Special issue of *Body and Society*, 6/3.

Macpherson C., 1962. *The political theory of possessive individualism. Hobbes to Locke*. Oxford University Press, Oxford.

Macpherson, C. 1978. *Property. Mainstream and critical positions*. University of Toronto Press, Toronto.

(MAFF) Ministry of Agriculture, Fisheries and Food, 1990. *Guidelines for the labelling of foods produced using genetic modification*. MAFF, London.

Mansell M., 1992. The court gives an inch and takes a mile. *Aboriginal Law Bulletin*, 2/57: 4–6.

Manton E., 1988. *Roman North Africa*. Seaby, London.

Margulis L. and R. Fester (eds), 1991. *Symbiosis as a source of evolutionary innovation*. The MIT Press, Cambridge, MA.

Margulis L. and K. Schwartz, 1982. *Five kingdoms*. Freeman, San Francisco, CA.

Markus A., 1994. *Australian race relations 1788–1933*. Allen and Unwin, Sydney.

Marsden T., A. Flynn and M. Harrison, 1999. *Consuming interests: the social provision of foods*. University College London Press, London.

Massey D., 1999a. Spaces of politics. In Massey D., J. Allen and P. Sarre (eds), *Human geography today*: 277–94. Polity Press, Oxford.

Massey D., 1999b. Space–time, 'science' and the relationship between physical geography and human geography. *Transactions of the Institute of British Geographers*, 24/3: 261–76.

Mathews F., 1991. *The ecological self*. Routledge, London.

Matuarana H. and F. Varela, 1992. *The tree of knowledge, the biological roots of human understanding*. Shambhala, Boston, MA.

Mayer S., J. Hill, R. Grove-White and B. Wynne, 1996. *Uncertainty, precaution and decision making: the release of genetically modified organisms in the environment*. Global Environmental Change Briefing No. 8. ESRC, Swindon.

Mazur B., 1995. Commercializing the products of plant biotechnology. *Trends in Biotechnology*, 13: 319–23.

McAfee K., 1998. Selling nature to save it: biodiversity and the rise of green developmentalism. *Society and Space*, 17: 133–54.

McCabe D., W. Swain, B. Martinell and P. Christou, 1988. Stable transformation of soybean (*Glycine max*) by particle acceleration. *Bio/Technology*, 6: 923–6.

McClintock A., 1994. *Imperial leather. Race, gender and sexuality in the colonial context*. Routledge, London.

McHugh J., 1992. What is the difference between a 'person' and a 'human being' within the law? *The Review of Politics*, 54/3: 445–61.

McKibben B., 1989. *The end of nature*. Anchor Books, New York.

McNeely J., K. Miller, W. Reid, R. Mittermeier and T. Werner, 1990. *Conserving the world's biodiversity*. IUCN, Switzerland.

Mercer D., 1997. Aboriginal self-determination and indigenous land title post-Mabo. *Political Geography*, 16/3: 189–212.

Merleau-Ponty M., 1962. *The phenomenology of perception* (trans. C. Smith). Humanities Press, New York.

Merleau-Ponty M., 1968. *The visible and the invisible* (trans. A. Lingis). Northwestern University Press, Evanston, IL.

Merleau-Ponty M., 1970. *Themes from lectures at the College de France 1952–60* (trans. J. O'Neill). Northwestern University Press, Evanston, IL.

Meyers D., 1994. *Subjection and subjectivity. Pyschoanalytic feminism and moral philosophy*. Routledge, London.

Michael M., 2000. *Reconnecting culture, technology and nature*. Routledge, London.

Midgley M., 1983. *Animals and why they matter*. University of Georgia Press, Athens GA.

Mill J.S. 1961 (1870). *Principles of political economy*. Augustus M. Kelley, New York.

Miller D., 1999. Risk, science and policy: definitional struggles, information management, the media and BSE. *Social Science and Medicine*, 49: 1239–55.

Miller H., 1999. A rational approach to labelling biotech-derived foods. *Science*, 284: 1471–2.

Minutes of Parliament (GB), 1834. *House of Commons debates on the South Australia Colonisation Bill*. Volumes 3: 2404–12 and 4: 2766–3190.

Mol A., 1999. A word and some questions. In Law J. and J. Hassard (eds), *Actor network theory and after*: 74–89. Basil Blackwell, Oxford.

Mol A. and J. Law, 1994. Regions, networks and fluids: anaemia and social topology. *Social Studies of Science*, 24: 641–71.

Monsanto, 1996. *The Soya Bean Information Centre*. press1.doc/1oct96

Monsanto, 2000. www.monsanto.com

Montgomery S., 1991. *Walking with the great apes: Jane Goodall, Dian Fossey, Biruté Galdikas*. Houghton Mifflin, Boston, MA.

Moraga C. and G. Anzaldua, 1981. *This bridge called my back*. Persephone, Watertown.

Morgan H., 1992. Exploration access and political power. *Mining Review*, 1 June: 26–33.

Morris M., 1988. Tooth and claw: tales of survival, and Crocodile Dundee. In *The pirate's fiancé*: 241–69. Verso, London.

Mouffe C., 1995. Post-Marxism: democracy and identity. *Society and Space*, 13/3: 259–65

Mulvany P., 1999. *Report on the Commission on Genetic Resources for Food and Agriculture 8th session*. http://www.fao/cgrfa8/htm

Munro N., 1997. Ideas of difference: stability, social spaces and the labour of division. In Hetherington K. and N. Munro (eds), *Ideas of difference*: 3–24. Basil Blackwell, Oxford.

Murdoch J., 1997a. Towards a geography of heterogenous associations. *Progress in Human Geography*, 21/3: 321–37.

Murdoch J., 1997b. Inhuman/nonhuman/human: actor-network theory and the potential for a non-dualistic and symmetrical perspective on nature and society. *Society and Space*, 15: 731–56.

Murdoch J., 1998. The spaces of actor-network theory. *Geoforum*, 29/4: 357–74.

Murdoch J., T. Marsden and J. Banks, 2000. Quality, nature and embeddedness. *Economic Geography*, 76/2: 107–25.

Murphy E., 1977. *Nature, bureaucracy and the rules of property*. North-Holland, New York.

Murphy W., 1994. As if: *camera juridica*. In Goodrich P. *et al.* (eds), *Politics, postmodernity and critical legal studies*: 69–106. Routledge, London.

Myerson G., 2000. *Donna Haraway and GM foods*. Postmodern encounters. Icon Books, Cambridge.

Naess A., 1989. *Ecology, community and lifestyle*. Cambridge University Press, Cambridge.

Nature, 1999. Science Policy Research Unit (SPRU) report on substantial equivalence, October.

Neeson J., 1993. *Customs in common. Common right, enclosure and social change in England 1700–1830*. Cambridge University Press, Cambridge.

New Scientist, 1999 (17 April). *Unpalatable truths*. www.newscientist.com/nsplus/insight/gmworld/gmfood.html

Noske B., 1989. *Humans and other animals: beyond the boundaries of anthropology*. Pluto Press, London.

Nottingham S., 1998. *Eat your genes. How genetically modified food is entering our diet*. Zed Books, London.

One Nation, 1999. http://www.onenation.com.au (established 1997).

O'Neill J., 1985. *Five Bodies*. Cornell University Press, Ithaca, NY.

O'Neill J., 1997. Time, narrative and environmental politics. In Gottlieb R. (ed.), *The ecological community*: 22–38. Routledge, London.

O'Riordan T., 1999. Dealing with scientific uncertainties. Paper presented at the Ditchley Foundation conference on 'Understanding, managing and presenting risk in public policy'. Mimeo.

(OST) Office of Science and Technology, 1999. *The advisory and regulatory framework for biotechnology: report from the Government's review*. May 1999. Cabinet Office and OST, London.

Ostrum E., J. Burger, C. Field, R. Norgaard and D. Policansky, 1999. Revisiting the commons: local lessons, global challenges. *Science*, 284: 278–88.

Pagden A., 1987. Dispossessing the barbarian: the language of Spanish Thomism and the debate over the property rights of the American Indians. In Pagden A. (ed.), *The languages of political theory in early modern Europe*: 79–119. Cambridge University Press, Cambridge.

Palast G., 1999. Dreaming in Monsanto. *Index on Censorship*, 3: 62–7.

(The) Parliament of the Commonwealth of Australia, 1993a. *Native Title Bill*. Commonwealth Government Printer, Canberra, Australia.

(The) Parliament of the Commonwealth of Australia, 1993b. *Native Title Act*. Commonwealth Government Printer, Canberra, Australia.

(The) Parliament of the Commonwealth of Australia, 1998. *Native Title Amendment Act*. Commonwealth Government Printer, Canberra, Australia.

Pateman C., 1989. *The disorder of women*. Stanford University Press, Stanford, CA.

Patton P., 1996. Sovereignty, law and difference in Australia: after the *Mabo* case. *Alternatives*, 21: 149–70.

Patton P., 2000. *Deleuze and the political*. Routledge, London.

Pearson N., 1993a. A troubling inheritance. Address to the Canberra Press Club, 10 November 1993. (Reprinted in *Race and Class*, 1994, 35/4: 1–9.)

Pearson N., 1993b. Reconciliation – to be or not to be. *Aboriginal Law Bulletin*, 3/61: 14–17.

Pearson N., 1993c. 204 years of invisible title. In Stephenson M. and S. Ratnapala (eds), *Mabo: a judicial revolution*: 75–95. University of Queensland Press, St Lucia, Queensland.

Pels P., 1998. The spirit of matter. On fetish, rarity, fact and fancy. In Spyer P. (ed.), *Border fetishisms*: 91–121. Routledge, London.

Pennisi E., 1999. Genomes reveal kin connections for whales and pumas. *Science*, 284: 2081.

Pesticides Safety Directorate, 1999. *Scientific review of the impact of Herbicide use on genetically modified crops*. PSD, Norwich.

Petchesky R., 1995. The body as property: a feminist re-vision. In Ginsberg F. and R. Rapp (eds), *Conceiving the new world order: the global politics of reproduction*: 387–406. University of California Press, Berkeley, CA.

Peterson N. and W. Sanders (eds), 1998. *Citzenship and indigenous Australians. Changing concepts and possibilities*. Cambridge University Press, Cambridge.

Philo C., 1995. Animals, geography and the city. *Society and Space*, 13/4: 655–81.

Philo C., P. Routledge and J. Sharp (eds), 1999. *Entanglements of power*. Routledge, London.

Philo C. and C. Wilbert (eds), 2000. *Animal spaces: beastly places*. Routledge, London.

Piercy M., 1992. *Body of glass*. Penguin books, Harmondsworth. (First published in the USA under the title *She, He and It*, 1991, Random House, New York.)

Pile S. and N. Thrift (eds), 1995. *Mapping the subject. Geographies of cultural transformation* (Introductory and concluding chapters by the editors). Routledge, London.

Plass P., 1995. *The game of death in Ancient Rome. Arena sport and political suicide*. University of Wisconsin Press, Madison, WI.

Pliny. *Natural History* (Penguin Classics edition published in 1991, trans. J. Healy). Penguin, Harmondsworth.

Plumwood V., 1993. *Feminism and the mastery of nature*. Routledge, London.

Polhill R. and J. Raven, 1981. *Advances in legume systematics*. Royal Botanic Gardens, Kew.

Poole R., 1991. *Morality and modernity*. Routledge, London.

Poole R., 2000. Justice or appropriation. Indigenous claims and liberal theory. *Radical Philosophy*, 101: 5–17.

Porter E., 1991. *Women and moral identity*. Allen and Unwin, Sydney.

Porter T., 1995. *Trust in numbers. The pursuit of objectivity in science and public life*. Princeton University Press, Princeton, NJ.

Posey D. and G. Dutfield, 1996. *Beyond intellectual property: toward traditional resource rights for indigenous peoples and local communities*. International Development Research Centre, Ottawa.

Probyn E., 1993. Technologizing the self. In Grossberg L. *et al.* (eds), *Cultural studies*: 501–11. Routledge, London.

Probyn E., 1996. *Outside belongings*. Routledge, London.

Probyn E., 1998. Mc-identities. Food and the familial citizen. *Theory, Culture and Society*, 15/1: 155–73.

Probyn E., 1999. Beyond food/sex. Eating and the ethics of existence. *Theory, Culture and Society*, 16/2: 215–28.

Rabb G. and T. Sullivan, 1995. Coordinating conservation: global networking for species survival. *Biodiversity and Conservation*, 4: 536–43.

Rabinow P., 1992a. Artificiality and enlightenment: from sociobiology to bio-sociality. In Kwinter J. and P. Crary (eds), *Incorporations*: 234–53. Zone Books, San Francisco, CA.

Rabinow P., 1992b. Severing the ties: fragmentation and dignity in late modernity. *Knowledge and Society*, 9: 160–87.

Rabinow P., 1996. *Making PCR. A story of biotechnology*. University of Chicago Press, Chicago, IL.

Raby P., 1996. *Bright paradise: Victorian scientific travellers*. Pimlico, London.

Radley A., 1995. The elusory body and social constructionist theory. *Body and Society*, 1/2: 3–23.

Raganarsdottir K. Vala, 2000. Environmental fate and toxicology of organophosphate pesticides. *Geochemistry and Health*, July.

Rajchman J., 2000. *The Deleuze connections*. The MIT Press, Cambridge, MA.

Ramakrishna K., 1992. North–South issues, the common heritage of mankind and global environmental change. In Rowlands I. and M. Greene (eds), *Global environmental change and international relations*: 145–68. Macmillan, London.

Ratzan S., 1998. *The mad cow crisis: health and the public good*. University College London Press, London.

Rawls J., 1971. *A theory of justice*. Oxford University Press, Oxford.

Reece B., 1987. Inventing Aborigines: assimilation of aborigines as a category into a national historiography. *Aboriginal History*, 11/1–2: 14–23.

Reeve A., 1986. *Property. Issues in political theory*. Macmillan, London.

Regan T. and P. Singer (eds), 1989 (2nd edition). *Animal rights and human obligations*. Prentice-Hall, Englewood Cliffs, NJ.

Report of the Royal Commission on Aboriginal Deaths in Custody, 1991. Commonwealth Printing Office, Canberra, Australia.

Revol B., 1995. Crocodile farming and conservation, the example of Zimbabwe. *Biodiversity and Conservation*, 4: 299–305.

Reynolds H., 1982 (2nd edition). *The other side of the frontier*. Penguin Australia, Ringwood, Victoria.

Reynolds H., 1988. *Aboriginal land rights in colonial Australia*. National Library of Australia, Occasional Lecture series no. 1, Canberra.

Reynolds H., 1992 (2nd edition). *The law of the land*. Penguin Books Australia, Ringwood, Victoria.

Reynolds H., 1996. *Aboriginal sovereignty. Three nations one Australia*. Allen and Unwin, Sydney.

Reynolds H., 1998. Sovereignty. In Peterson N. and W. Sanders (eds), *Citizenship and indigenous Australians. Changing conceptions and possibilities*: 208–15. Cambridge University Press, Cambridge.

Rheinberger H.-J., 1997. *Towards a history of epistemic things: synthesizing proteins in the test tube*. Stanford University Press, Stanford, CA.

Ridley R. and H. Baker, 1998. *Fatal protein: the story of CJD, BSE and other prion diseases*. Oxford University Press, Oxford.

Rieseberg L. and N. Ellstrand, 1993. What can molecular and morphological markers tell us about plant hybridization. *Critical Reviews in Plant Sciences*, 12/3: 213–41.

Ritvo H., 1995. Border trouble: shifting lines of demarcation between animals and humans. Special issue 'In the company of animals'. *Social Research*, 62: 481–500.

Ritvo H., 1997. Flesh made word. In *The Platypus and the mermaid and other figments of the classifying imagination*: 51–84. Harvard University Press, Cambridge, MA.

Roberts R. and J. Emel, 1992. Uneven development and the tragedy of the commons: competing images for nature–society relations. *Economic Geography*, 68: 249–71.

Robertson G., M. Mash, L. Tickner, J. Bird, B. Curtis and T. Putnam (eds), 1996. *FutureNatural: nature/science/culture*. Routledge, London.

Rodman G., 1993. Making a better mystery out of history. *Meanjin*, 2/winter: 295–312.

Rogoff I., 2000. Mapping. In *terra infirma: geography's visual culture*: 73–111. Routledge, London.

Rose C., 1986. The comedy of the commons. Culture, commerce and inherently public rights. *University of Chicago Law Review*, 53/3: 711–811.

Rose D. Bird, 1984. The saga of Captain Cook: morality in Aboriginal and European law. *Australian Aboriginal Studies*, 2: 24–39.

Rose S., 1997. *Lifelines. Biology, freedom, determinism*. Penguin, Harmondsworth.

Rost T., M. Barbour, C. Stocking and T. Murphy, 1998. *Plant biology*. Wadsworth, Belmont, CA.

Rothman H., P. Glasner and C. Adams, 1996. Proteins, plants and currents: rediscovering science in Britain. In Irwin A. and B. Wynne (eds), *Misunderstanding science? The public reconstruction of science and technology*: 191–212. Cambridge University Press, Cambridge.

Rowse T., 1993a. Giving ground. *Arena Magazine*, Oct.–Dec.: 6–19.

Rowse T., 1993b. *After Mabo. Interpreting indigenous traditions*. Melbourne University Press, Melbourne.

Rowse T., 1993c. Mabo and moral anxiety. *Meanjin*, 52/2: 229–52.

(The) Royal Academy of Arts, 1999. *Joseph Beuys: the secret block for a secret person in Ireland*. Royal Academy Publications, London.

Ruddick S., 1989. *Maternal thinking: towards a politics of peace*. Ballantine Books, New York.

Ruiz L., 1991. After national democracy: radical democratic politics at the edge of modernity. *Alternatives*, 16: 161–200.

Ryan A., 1984. *Property and political theory*. Basil Blackwell, Oxford.

Sachs W., 1993. One-world. *The development dictionary*: 102–15. Zed Books, London.

Sachs W., 1994. *Global ecology*. Zed Books, London.

Sagan D., 1992. *Metametazoa: biology and multiplicity*. In Kwinter J. and P. Crary (eds), *Incorporations*: 362–85. Zone Books, San Francisco, CA.

Sandel M., 1982. *Liberalism and the limits of justice*. Cambridge University Press, Cambridge.

Sarkar S., 1998. *Genetics and reductionism*. Cambridge University Press, Cambridge.

Scarre C., 1995. *The Penguin atlas of Ancient Rome*. Penguin, London.

Schatzki T., K. Knorr-Cetina and E. von Saviny (eds), 2000. *The practice turn in contemporary theory*. Routledge, London.

Schmidt K., 1995. Whatever happened to the gene revolution? *New Scientist*, 7 January: 21–5.

Scholtmeijer M., 1997. What is 'human'? Metaphysics and zoontology in Flaubert and Kafka. In Ham J. and M. Senior (eds), *Animal acts*: 127–44. Routledge, London.

Schultz A., F. Wengenmayer and H. Goodman, 1990. Genetic engineering of herbicide resistance in higher plants. *Critical Reviews in Plant Sciences*, 9/1: 1–15.

Scott D., 1940. Taking possession of Australia – the doctrine of *Terra nullius*. *Royal Australian Historical Society Journal and Proceedings*, 26/1: 5–12.

Sedjo R., 1992. Property rights, genetic resources and biotechnological change. *Journal of Law and Economics*, 35: 199–213.

Senior M., 1997. 'When the beasts spoke'. Animal speech and classical reason in Descartes and La Fontaine. In Ham J. and M. Senior (eds), *Animal acts*: 61–84. Routledge, London.

Serres M., 1975. *Feux et signaux de brume*. Grasset, Paris.

Serres M., 1985. *Les cinq sens*. Grasset, Paris.

Serres M., 1991. *Rome. The book of foundations* (trans. F. McCarren). Stanford University Press, Stanford, CA.

Serres M., 1993. *Angels: a modern myth* (trans. F. Cowper). Flammarion, Paris.

Serres M., 1995. *The natural contract* (trans. E. MacArthur and W. Paulson). Michigan University Press, Ann Arbor, MI.

Serres M. and B. Latour, 1995. *Conversations on science, culture and time* (trans. R. Lapidus). University of Michagan Press, Ann Arbor, MI.

Shapiro I., 1991. Resources, capacities and ownership. The workmanship ideal and distributive justice. *Political Theory*, 19/1: 47–72.

Shapiro M., 1991. Sovereignty and exchange in the orders of modernity. *Alternatives*, 16/4: 447–77.

Sheehan J. and M. Sosna, 1991. *The boundaries of humanity. Humans, animals, machines.* University of California Press, Berkeley, CA.

Sheets-Johnstone M., 1992. Corporeal archetypes and power: preliminary clarifications and considerations of sex. *Hypatia*, 7/3: 39–76.

Sheets-Pyenson S., 1988. *Cathedrals of science. The development of colonial natural history museums during the late 19th century.* McGill-Queens University Press, Montreal.

Shepard P., 1997. *The others: how animals made us human.* Island Press, Washington, DC.

Shields R., 1992. A truant proximity: presence and absence in the space of modernity. *Society and Space*, 10: 181–98.

Shields R., 1997. Flow. *Space and Culture*, 1: 1–7.

Shiva V., 1993. *Monocultures of the mind. Perspectives of biodiversity and biotechnology.* Third World Network Publishers, Penang, Malaysia.

Shotter J., 1993. *Cultural politics of everyday life: social constructionism, rhetoric and knowing of the third kind.* Open University Press, Buckingham.

Shusterman P., 2000. *Performing live: Aesthetic alternatives as for the ends of art.* Cornell University Press, Ithaca, NY.

Simmonds N., 1979. *Principles of crop improvement.* Longman, New York.

Simmons I., 1996 (2nd edition). *Changing the face of the earth: culture, environment, history.* Basil Blackwell, Oxford.

Simpson A., 1986 (2nd edition). *A history of land law.* Hambledon, London.

Simpson A., 1987. *Legal theory and legal history: essays on the common law.* Hambledon, London.

Simpson G., 1993. *Mabo*, international law, *terra nullius* and the stories of settlement: an unresolved jurisprudence. *Melbourne University Law Review*, 19: 195–210.

Smith A., 1986 (1st published in 1776). *An inquiry into the nature and causes of the wealth of nations.* Penguin, London.

Smith N., 1990. (1st edition 1984). *Uneven development.* Basil Blackwell, Oxford.

Soil Association, 1999. *The state of organics in the UK.* Soil Association, Bristol.

Soja E., 1996. *Thirdspaces.* Basil Blackwell, Oxford.

Soper K., 1995. *What is nature?* Basil Blackwell, Oxford.

Soulé M. and G. Lease (eds), 1995. *Reinventing nature? Responses to postmodern deconstruction.* Island Press, Washington, DC.

Southwood Report. 1989. Report of the Working Party on Bovine Spongiform Encephalopathy. www.bse.org.uk/files/ib/ibd1/tab2.pdf

Squadrito K., 1979. Locke's view of dominium. *Environmental Ethics*, 1: 255–62.

Squires J. (ed.), 1993. *Principled positions: postmodernism and the rediscovery of value.* Routledge, London.

Spary E., 2000. *Utopia's garden. French natural history from old regime to revolution.* University of Chicago Press, Chicago, IL.

Star S. Leigh, 1991a. Power, technology and the phenomenology of conventions: on being allergic to onions. In Law J. (ed.), *A sociology of monsters*: 26–56. Basil Blackwell, Oxford.

Star S. Leigh, 1991b. The sociology of an invisible: the primacy of work in the writings of Anselm Strauss. In Maines D. (ed.), *Social organisation and social processes: essays in honour of Anselm L. Strauss*: 98–126. Aldine de Gruyter, Hawthorne, NY.

Stengers I., 1996. *Cosmopolitiques* (2 volumes). La Decouverte, Paris.

Stengers I., 1997. *Power and invention. Situating science* (trans. P. Bains). University of Minnesota Press, Minneapolis, MN.

Stephenson M. and S. Ratnapala (eds), 1993. *Mabo: a judicial revolution. The aboriginal land rights decision and its impact on Australian law*. University of Queensland Press, St Lucia, Queensland.

Sterchi B., 1988. *The cow* (trans. M. Hofmann). Faber and Faber, London.

Stone N., 1995. Sweeping patents put biotchnology companies on the warpath. *Science*, 268: 656–8.

Strabo, 1916. *The geography of Strabo* (T. Bohn's Classical Library edition, trans. H. Hamilton and W. Falconer). G. Bell & Sons, London.

Strathern M., 1996. Cutting the network. *Journal of the Royal Anthropological Society* (N.S.), 2: 517–35.

Strathern M., 1998. Divisions of interest and languages of ownership. In Haan C. (ed.), *Property relations*: 214–32. Cambridge University Press, Cambridge.

Strathern M., 1999a. What is intellectual property after? In Law J. and J. Hassard (eds), *Actor-network theory and after*: 156–80. Blackwell, Oxford.

Strathern M., 1999b. *Property, substance and effect. Anthropological essays on persons and things*. Athlone Press, London.

Systematic Zoology, 1959. Symposium on Linnaeus and nomenclatorial codes. *Systematic Zoology*, 8: 1–47.

Takacs D., 1996. *The idea of biodiversity. Philosophies of paradise*. Johns Hopkins University Press, Baltimore, MD.

Taussig M., 1993. *Mimesis and alterity*. Routledge, London.

Terkel A., 1996. The European EEP list for the African Elephant, *Loxodonta africana*. Unpublished manuscript, Zoological Center, Tel Aviv-Ramat Gan.

Thompson E.P., 1991. *Customs in common*. Penguin, London.

Thorbjarnarson J., 1999. Crocodile tears and skins: international trade, economic constraints, and limits to the sustainable use of crocodilians. *Conservation Biology*, 13/3: 465–70.

Thorne L., 1998. Kangaroos: the non-issue. *Society and Animals*, 6/2: 167–82.

Thrift N., 1991. Over wordy worlds? Thoughts and worries. In Philo C. (ed.), *New words, new worlds: reconceptualising social and cultural geography*: 144–8. Social and Cultural Geography Study Group, Institute of British Geographers, Lampeter.

Thrift N., 1996. *Spatial Formations*. Sage, London.

Thrift N., 1999. Steps to an ecology of place. In Massey D., J. Allen and P. Sarre (eds), *Human geography today*: 295–322. Polity Press, Cambridge.

Thrift N., 2000a. Still life in nearly present time: the object of nature. *Body and Society*, 6/3–4: 34–57.

Thrift N., 2000b. Afterwords. *Society and Space*, 18/2: 213–56.

Thrift N. and K. Olds, 1996. Refiguring the economic in economic geography. *Progress in Human Geography*, 20: 311–37.

Time Magazine, 1996. The battle of the bean genes. *Time Magazine*, 28 October: 48–9.

Toynbee J.M.C., 1973. *Animals in Roman life and art.* Thames and Hudson, London.

Tribe K., 1978. *Land, labour and economic discourse.* RKP, London.

Tuck R., 1979. *Natural right theories.* Cambridge University Press, Cambridge.

Tudge C., 1993. *The engineer in the garden. Genetics: from the idea of heredity to the creation of life.* Pimlico, London.

Tully J., 1980. *Discourse on property. John Locke and his adversaries.* Cambridge University Press, Cambridge.

Tully J., 1993. *An approach to political philosophy: Locke in contexts.* Cambridge University Press, Cambridge.

Tully J., 1995. *Strange multiplicity. Constitutionalism in an age of diversity.* Cambridge University Press, Cambridge.

Turney J., 1998. *Frankenstein's footsteps. Science, genetics and popular culture.* Yale University Press, New Haven, CT.

(UPOV) Union pour la Protection des Obtentions Végétales, 1972. *Convention Internationale pour la protection des obtentions végétales.* Union pour la Protection des Obtentions Végétales, Geneva.

Valceschini E., 1998. *L'étiquetage obligatoire des aliments est-il la meilleure solution pour les consommateurs?* Paper for conference on 'Environnement et alimentation', INRA. Mimeo.

Varela F., 1992. The re-enchantment of the concrete. In Crary J. and S. Kwinter (eds), *Incorporations*: 320–38. Zone Books, San Francisco, CA.

Varela F., 1999. *Ethical know-how.* Stanford University Press, Stanford, CA.

Varela F., E. Thompson and E. Rosch, 1991. *The embodied mind: cognitive science and human experience.* The MIT Press, Cambridge, MA.

de Vattel E., 1916 (1st published in English in 1760). *The law of nations or the principles of natural law* (3 volumes). Carnegie Institute, Washington, DC.

Ville G., 1981. *La gladiature en occident des origines a la mort de Domitien.* Ecole Française de Rome, Rome.

Visvanathan S., 1991. Mrs Brundtland's disenchanted world. *Alternatives*, 16: 377–84.

Vogel U., 1988. When the earth belonged to all: the land question in eighteenth-century justifications of private property. *Political Studies*, 36: 102–22.

von Uexküll J., 1992 (1st published in 1934). A stroll through the worlds of animals and men: a picture book of invisible worlds. *Semiotica*, 89/4: 319–91.

von Weisäcker C., 1993. Competing notions of biodiversity. In Sachs W. (ed.), *Global ecology*: 117–31. Zed Books, London.

von Weisäcker H., 1994. *Earth politics.* Zed Books, London.

Wake C. Stanisland, 1872. The mental characteristics of primitive man as exemplified by the Australian Aborigines. *Royal Anthropological Institute of Great Britain and Ireland*, 1: 72–85.

Walden R. and R. Wingender, 1995. Gene-transfer and plant-regeneration techniques. *Trends in Biotechnology*, 13: 324–30.

Walker R., 1991. On the spatio-temporal conditions of democratic practice. *Alternatives*, 16: 243–62.

Watson J., M. Gilman, J. Witkowski and M. Zoller, 1992 (2nd edition). *Recombinant DNA.* Scientific American Books, New York.

(WCED) World Commission on Environment and Development, 1987. *Our common future.* Oxford University Press, Oxford.

Webster G. and B. Goodwin, 1996. *Form and transformation. Generative and relational principles of biology.* Cambridge University Press, Cambridge.

Weismann A., 1892. *Essays upon heredity and kindred biological problems* (2 volumes edited by Poulton E. and A. Shipley). Clarendon Press, Oxford.

Weiss G., 1999. *Body images. Embodiment as intercorporeality.* Routledge, London.

Wells H.G., J. Huxley and G.P. Wells, 1931 (1st edition). *The science of life.* Cassell, London.

Welton D. (ed.), 1999. *The body.* Basil Blackwell, Oxford.

Whatmore S. and L. Thorne, 1997. Nourishing networks: alternative geographies of food. In Goodman D., and M. Watts (eds), *Globalising food: agrarian questions and global restructuring*: 287–304. Routledge, London.

White H., 1978. The forms of wildness: archaeology of an idea. In *Tropics of discourse: essays in cultural criticism*: 150–82. Johns Hopkins University Press, Baltimore, MD.

Whitehead A. North, 1929. *Process and reality. An essay in cosmology.* Cambridge University Press, Cambridge.

Whitt L. and J. Slack, 1994. Communities, environments and cultural studies. *Cultural Studies,* 8/1: 5–31.

Wiedemann T., 1992. *Emperors and gladiators.* Routledge, London.

Wiel S., 1934. Natural communism: air, water, oil, sea, and seashore. *Harvard Law Review,* 47: 425–57.

Willems-Braun B., 1997. Buried epistemologies: the politics of nature in (post) colonial British Columbia. *Annals of the Association of American Geographers,* 87/1: 3–31.

Wilkes G., 1983. Current status of crop germplasm. *Critical Reviews in Plant Sciences,* 1/2: 133–81.

Williams R., 1989. *Resources of hope.* Verso, London.

Wilmut I., A. Schnieke, J. McWhir, A. Kind and K. Campbell, 1997. Viable offspring derived from fetal and adult mammalian cells. *Nature,* 385: 810–13.

Winterson J., 1997. *Gut symmetries.* Alfred Knopf, New York.

Wise S., 2000. *Rattling the cage. Towards legal rights for animals.* Profile Books, London.

Woese C., O. Kandler and M. Wheelis, 1990. Towards a natural system of organisms: proposal for the domains Archaea, Bacteria and Eukarya. *Proceedings of the National Academy of Sciences,* 87: 4576–9.

Wolch J. and J. Emel, 1995. Bringing the animals back in. Guest editorial to special issue of *Society and Space,* 13/6: 632–6.

Wolch J. and J. Emel (eds), 1998. *Animal geographies.* Verso, London.

Wolch J., K. Wesk and T. Gaines, 1995. Transpecies urban theory. *Society and Space,* 13/4: 735–60.

Wolfe C., 1998. *Critical environments. Postmodern theory and the pragmatics of the 'outside'.* Minnesota University Press, Minneapolis, MN.

Woodmansee M., 1984. The genius and the copyright: economic and social conditions of the emergence of the 'Author'. *Eighteenth-Century Studies,* 17/4: 425–48.

Xenos N., 1989. *Scarcity and possession.* Routledge, London.

Yapa L., 1993. What are improved seeds? An epistemology of the Green Revolution. *Economic Geography*, 69/3: 254–73.

Young I., 1989. Polity and group difference: a critique of the ideal of universal citizenship. *Ethics*, 99: 250–74.

Zuckerman L., 1998. *The potato. How the humble spud rescued the western world.* Faber and Faber, Boston, MA.

Index of Names

Page numbers referring to the names of authors who have been quoted or discussed are in **bold** type. Authors who have been only referred to are in normal type.

Subject Index

Definitions of terms in **bold**
In this index the abbreviation PGR stands for Plant Genetic Resources.

Printed in the United States
135711LV00002B/21/A